12/90

D0635792

SOWING
THE
WIND

SOWING THE WIND

—————— • ——————

*Reflections on the
Earth's Atmosphere*

LOUISE B. YOUNG

PRENTICE
HALL
PRESS

New York • London • Toronto • Sydney • Tokyo • Singapore

Prentice Hall Press
15 Columbus Circle
New York, New York 10023

PRENTICE HALL PRESS and colophons are registered trademarks
of Simon & Schuster, Inc.

Library of Congress Cataloging-in-Publication Data

Young, Louise B.
    Sowing the wind : reflections on the earth's atmosphere / Louise
B. Young
        p. cm.
    ISBN 0-13-083510-2 : $17.95
    1. Air—Pollution. 2. Atmosphere. I. Title.
TD883.Y68   1990
363.7'392—dc20   89-23145
                                                           CIP

Designed by
Irving Perkins Associates

Manufactured in the United States of America

10   9   8   7   6   5   4   3   2   1

First Edition

To the men and women who took the first steps into space and, looking back, saw our planet whole—"that beautiful, warm living object, so fragile, so delicate," exclaimed James Irwin (USA), "seeing this has to change a man." . . . "It was small, light blue, and so touchingly alone," said Alexei Leonov (USSR). "Our home that must be defended like a holy relic."

Let us pray this new vision will change us all.

Acknowledgments

In writing this book I have drawn material from a number of different fields of science and have needed the help of experts in these rapidly advancing fields of knowledge. I have been fortunate in having the cooperation of a number of distinguished scientists who have been generous with their time and expertise. They have read and commented on various pieces of the manuscript, offering valuable suggestions and criticisms. For such very constructive help I am indebted to Cheves T. Walling, Professor of Chemistry, University of Utah; Roscoe R. Braham, Jr., Professor of Meteorology, University of Chicago; Thomas Murphy, Professor of Chemistry, DePaul University; Reid A. Bryson, Professor of Meteorology and Climatology, Center of Climatic Research, University of Wisconsin; Eugene Parker, Professor of Physics and Astronomy, University of Chicago; J. John Sepkowski, Jr., Professor of Paleontology, University of Chicago; and Dr. William R. Moomaw, Director of the Climate, Energy, and Pollution Program at World Resources Institute.

Many other scientists have been helpful in describing their fields of special interest, in answering my questions, and in providing valuable source materials. Among those I especially want to thank V. Ramanathan and John Frederick, both Professors of Atmospheric Sciences at the University of Chicago, and Dr. Susan Solomon, National Oceanic and Atmospheric Administration.

In spite of the expert help and advice I have received, I am very much aware of my own obligations in presenting the scientific material fairly and correctly. Final responsibility for the accuracy of the information and for the interpretation of the facts is mine alone.

L. B. Y.

Wilmette, Illinois
October 1989

Contents

Introduction

Since the dawn of the Industrial Era human beings have been sowing the winds of the Earth with the seeds of change. We have emptied the refuse of our factories into the finely articulated layer of blue vapor that enfolds our planet. We have added strange new man-made chemicals to this sensitive medium, and no one knows exactly how great the ultimate consequences of these changes may be.

The signs of our interference with nature have actually been present for several decades, but until just recently they have not been fully recognized. Warnings issued by atmospheric scientists have been ignored or passed over very lightly. Now, in the wake of one very hot summer (1988) and one warm decade, a sudden surge of interest has focused the attention of the whole world on these problems. It is easy to understand how environmentalists, swept up in this new wave of public interest, have been anxious to take advantage of the opportunity it offers to achieve action on issues they have long believed to be important.

It is prudent to remember, however, that fears generated in an atmosphere of crisis may be easily forgotten if predictions of disaster are not substantiated in the next year or two—or possibly three. Although dramatic statements may seize the attention of the public for a brief moment, they will not sustain the long-term commitment that is needed to deal with problems that have been a hundred years in the making. The experts that overstate their case run the risk of losing their constituents, and the entire environmental cause might suffer a serious setback.

A situation of this kind developed in the Chicago area a few years ago. In 1985, the water levels in the Great Lakes had reached quite high levels. But there was no immediate cause for alarm because, historically, these levels have risen and receded approximately every ten or twelve years. Although this has not been an exact cycle, the changes have occurred about eight times during this century.

In the fall of 1986, unusually heavy rains fell in the watershed of the Great Lakes, raising the water to the highest levels ever recorded. Lake Michigan, for example, reached three-and-one-half feet above average high levels, as measured by the Army Corps of Engineers during the past 126 years.* Record precipitation continued throughout that winter and the following year. Storm waves scoured out beaches, damaged sea walls, and pounded against bluffs. Towering crests hurled spray against windows and balconies of lakefront buildings. In winter the spray turned to ice, shattering windows and breaking the branches of trees planted nearby.

Panic seized the cities and communities along the shores of Lake Michigan. Almost every day, newspapers and television programs carried new disaster stories with pictures of eroded bluffs and houses slumped into the lake. Experts were called in to evaluate the problem. How long would it last? How high will the waters rise? Their statements were not reassuring; in fact, they fed the panic.

It happened that new scientific information on lake levels had recently been discovered. The results of a geological study, published just a few months earlier, provided evidence of ancient water level fluctuations in Lake Michigan. Radiocarbon dating showed that during the last 2,000 years the levels of the lake had risen and fallen in very long

* The reference notes for this book can be found in a section beginning on page 187. They are arranged by page number with a short phrase identifying the subject matter for each note. No reference numbers will appear in the text.

rhythms—on the order of 500 years, and high levels lasted for periods as long as a century.

From the scientists' point of view this was fascinating information, and they emphasized it in their interviews. Little or nothing was said about the more recent history of short-term fluctuations of a smaller magnitude.

The message that came across to the people of Chicago and the other lakeshore communities was that Lake Michigan would continue to rise, reaching levels three to five feet higher than the current, already alarmingly destructive levels. No relief could be expected for at least a decade. The value of lake-front property fell precipitously and owners rushed to install protective walls and piles of rocks. Calls for government action to alleviate the problem resulted in many conferences and a broad range of proposals for diverting water from the Great Lakes as well as laws restricting the use of lakeshore property.

Then, in 1987, water levels began to recede, and this trend continued throughout the next two years. By the summer of 1989, the levels had declined to a point several inches *below* the average levels for this century. The people who lived near the lake smiled and went back out to enjoy the beaches. So much for the opinion of experts!

But, in fact, the water-level fluctuations of the Great Lakes remain a long-term problem to be addressed. Other high-level episodes will probably occur. Low levels are also troublesome because of the obvious disadvantages to shipping. It is a pity that the current loss of credibility has now made it more difficult to achieve public and political support for action on controlling the extremes of water levels and the use of the shoreline. A more moderate response to the crisis could have resulted in a sustained effort that would have led to constructive measures.

I am deeply concerned that the crisis psychology generated by some of the more extreme statements in the press dealing with atmospheric issues, such as global warming, could result in a similar loss of credibility. The public has been told that we have entered a long period of rising temperatures throughout the world; unless drastic action is taken, the warming will continue at an ever accelerating rate. So what will happen if this prediction is not borne out in the immediate future? Two cold years coming close together could be enough to prick the bubble of public interest. This danger could be avoided were the experts to describe, dispassionately, the state of scientific knowledge on these subjects, being careful to assess the various factors that might *change* the prognosis. People in general believe that scientists will give them the whole truth, and this is a trust that should not be violated. Failure to honor this trust may eventually cause a reaction against the environmental movement.

Those who have worked for many years in this cause, as I have, would be sad indeed to see this happen. There is so much that needs to be done to prevent accelerating deterioration of our environment, and this effort demands an understanding of all the facts as they are revealed each step of the way. Action undertaken in the sudden glare of a crisis will not provide a long-term solution to problems that have been slowly building for many decades, just as a crash diet cannot correct an overweight problem of many years standing. What is needed is a change in daily habits, a *sustained* commitment to the health of our planet in the face of increasing population pressure and rising expectations of all humanity.

— • —

In this book I attempt to present the interested layman with a balanced picture of the scientific knowledge on the issues of greenhouse warming, acid rain, and the depletion of the ozone layer. Because I believe it is important to have as a background some understanding of the structure of our atmosphere and a knowledge of the natural processes that affect the weather, the four chapters in Part I cover these general subjects. They also give a brief review of the history of climate changes, so that the present situation can be seen in perspective. The chapters in Part II deal with the specific problems.

In a book of this length it is impossible to cover all the factors that are important in maintaining the quality of life on Planet Earth: toxic wastes, oil spills, pesticides, and the increase of human population. These and other important subjects I have had to omit. In this book I have concentrated on the atmospheric problems that have the greatest world-wide significance. The international aspect of these issues presents us with a unique opportunity. For the first time in human history we find that we must cooperate with all the other people on Earth to fight a common enemy. The enemy turns out to be ourselves. As Arthur Eddington said (in another context): "We have found a strange footprint on the shores of the unknown. We have devised profound theories, one after another, to account for its origin. At last, we have succeeded in reconstructing the creature that made the footprint. And lo! It is our own."

PART I

·

THIS EXCELLENT CANOPY

1

— • —

Exploring the Troposphere

Cirrus clouds are composed of cascades of ice crystals. (Courtesy of NASA.)

The Space Age has changed our perspective on our home in the universe—this small planet spinning on the periphery of the Milky Way, 25,000 light-years from its center, and 14 billion light-years from the farthest galaxies on the edge of the observable cosmos. When man broke free from the bonds of gravity that had held him close to the surface of the Earth, he was able to look back, for the first time, and see this luminous bubble of matter floating in the blackness of space. He could see the thin, blue layer of vapor that surrounds the planet. He saw that it looks too insubstantial to hold any significance in this vast abyss. Yet to us on Earth it is everything. It is the limitless blue sky in which the eagle soars; it is the wind that rocks the robin's nest; and it is the hoarfrost on grass, the morning dew. It is the softness of spring and the bite of winter. Indeed it is the very seasons themselves.

This fragile envelope of atmosphere nourishes and protects us. It shelters us from the extreme temperature differences of space by catching and diffusing the light of the Sun, filtering out the most damaging rays. It absorbs, alters, and recirculates the many elements essential for life—the oxygen for animals to breathe and carbon dioxide for all the green, growing things. From the seas and lakes it draws up water, distills it, purifies it, and distributes it again across the planet's surface.

Not only does the atmosphere alter the conditions on Earth and make life possible, it is also formed and changed by them. It receives the fiery breath of the planet itself that pours forth from volcanoes and the hot plumes of effluent that boil up from the bottom of the sea. All forms of life contribute their share to the constantly changing composition of this thin envelope of vapor. Man, too, with his fur-

5

naces and his agriculture alters the atmosphere just as surely as the conditions of his life are created by it.

Our atmosphere is a medium rich in variety, subtle in response to the many influences that impinge upon it. We cannot hope to preserve the quality of this dynamic medium or make a prognosis for its future without considering all its complexities.

For most of the time that mankind has existed on this planet, the nature of the atmosphere has been a mystery. Its moods, sometimes delightful, sometimes cruel and life-threatening, were attributed to gods who must be propitiated—gods who could withhold the rain, hurl thunderbolts, and set fires that rage across the prairies and forests of the Earth. Real knowledge about the atmosphere did not begin until just a few centuries ago when human beings began to explore this unknown realm.

The first steps were taken by mountain climbers, men who were not seeking knowledge but a challenge and an adventure. Up to heights of about two miles, mountaineering provides an exhilarating experience. Mountain climbers have often spoken of the euphoria that sweeps over them when they stand on a mountain top surrounded by air so crystalline that the whole sky seems to ring and the air to be shattered by the Sun's rays into showers of light. The Earth lies spread beneath their feet in a great circle, with every field and woodland, every town and farmhouse delineated with amazing clarity. Only out toward the expanded horizons are the far hills softened with a faint purple haze. But when human beings climb higher, they encounter more and more unfavorable conditions. They begin to suffer headaches and nausea, and have a feeling of heaviness and lassitude. The reasons for these symptoms were not well understood until men had ventured much farther into the atmosphere—had scaled Mt. Everest (about five-and-a-half miles high) and had flown to even greater altitudes in balloons. As in all exploration of the unknown, the knowledge was acquired at considerable risk of life and produced many hair-raising adventures.

By the end of the eighteenth century it was well known that the air becomes thinner and lighter as one moves away from the Earth's surface. The barometer had been invented, and it was possible to measure the gradual decrease in pressure with increasing altitude. It was also known that hot air rises, a fact that could be observed every day by watching the plume of smoke from a fire or feeling the updraft over a lighted candle.

The reason for this phenomenon can be easily understood by remembering that air (or anything in a gaseous state) contains billions of molecules all free to move independently and all moving in a random way at high speeds. They constantly collide with each other and with anything that happens to be in their way. The average velocity of their motion squared is proportional to the quantity we measure as temperature. Each molecule is subject to the attraction of gravity but resists this force by virtue of its speed. The faster the molecules are moving, the more they tend to spread out, and if they are in a flexible container such as a balloon, the greater the speed of the molecules, the more inflated the balloon becomes. The hot air in the balloon is lighter than the air it displaces and, therefore, the balloon rises.

This fact was first exploited in the late eighteenth century by two French brothers, Étienne and Joseph Montgolfier. They built a balloon of linen that was lined with paper. It was filled with hot air, and two volunteer pilots rode in a little open wicker basket, stoking a straw fire to maintain the temperature of the air in the balloon. During the brief flight, sparks from the fire burned holes in the balloon, but the men managed to land safely after traveling eight miles and reaching an altitude of 3,000 feet.

Another method of obtaining buoyancy was the use of

hydrogen instead of hot air. Hydrogen is lighter than air—almost sixteen times—but it can be dangerous, too. Any spark can touch off a fire and, when mixed with air, a violent explosion. Coal gas was a third option. This gas, made from the distillation of coal, is composed chiefly of hydrogen and methane (a compound of hydrogen and carbon), and it is also lighter than air.

Throughout the next century these three technologies were all used to carry men on extraordinary expeditions into the new frontier. They crossed the Alps and the English Channel, spied on enemy positions during the Civil War, and carried mail by air more than 600 miles. In 1870, during the German siege of Paris, a flotilla of sixty-six balloons transported messages from the besieged city over the German lines. Three million dispatches were delivered in this way and reconnaissance information brought back to the city helped to shorten the siege.

But not all the adventures had a happy ending; several resulted in disaster. One of the most tragic expeditions was launched by Salomon Auguste Andrée, a Swedish explorer. In a hydrogen balloon, Andrée and two companions set off from Dane's Island in the Barents Sea with the objective of reaching the North Pole. More than three decades later the frozen bodies of the three men and remains of their equipment were discovered thirty-five miles east of their take-off point at Spitsbergen.

During the early years of ballooning, it was not known just how far an unprotected man could safely venture into this ocean of air. Several of the early aeronauts reported strange and alarming symptoms when their balloons reached altitudes so great that ice began to form on their clothes and bodies. They had difficulty breathing, felt oppression on the chest, experienced vomiting and nosebleeds. Scientists wondered whether these effects could be caused by insufficient oxygen because the air is thinner at these altitudes. They knew that air is composed of several different elements. Nitrogen is the dominant factor—78 percent; oxygen is a little less than 21 percent, and then there are a number of trace gases. But molecules of oxygen are slightly heavier than those of nitrogen and are therefore more strongly attracted by gravity to the Earth. So oxygen might thin out more rapidly than nitrogen with increasing altitude.

In 1862, the British Association for the Advancement of Science decided to send one of their colleagues up to investigate these problems and make scientific measurements. James Glaisher, a scientist of some reputation who had founded the British Meteorological Society and helped to set up the Aeronautical Society of Great Britain, volunteered for the flight. He was a dignified gentleman, fifty-three years old, with a bald head and mutton-chop whiskers. Glaisher was joined in the experiment by Henry Coxwell, who, at forty-three, had become known as the best aeronaut in England.

Coxwell built the balloon. It was fifty-three feet in diameter and designed to be filled with coal gas. Like most aerostats, it was equipped with a valve that was controlled from the basket by a cord. In this way it could be opened to allow the gas to escape when the crew wanted to descend. The expedition carried a number of scientific instruments for measuring altitude, humidity, temperature, the Earth's magnetic field, and the solar spectrum. Sandbags were carried for ballast as well as a crate of pigeons for testing the effect of high altitude on other living things.

On the morning of September 5, 1862, Coxwell and Glaisher readied their balloon for flight. The weather was threatening, but they dared not delay their departure again. It had already been postponed three times, and the representatives of the Balloon Committee of the British Association for the Advancement of Science were growing impatient. Black clouds loomed overhead and a very strong ground wind was blowing when Glaisher and Coxwell climbed into the basket with their load of fragile instruments and the crate of pigeons. No sooner had they cut loose than a driving rain began and the balloon was bounced roughly across the launching field. Then suddenly it was seized by a current of air and carried steeply upward into the clouds. The two men were drenched by torrents of rain, buffeted by gusts of wind. Emerging briefly into a brighter sky, they saw a gigantic yellow thunderhead above them. In another instant they were sucked up into it. Their unwieldy craft spun violently in the vortex of a giant maelstrom. Thunder and lightning crashed about them as they crouched against the sides of their basket, trying to shield the bundle of delicate scientific instruments with their bodies.

Then, all at once, the sky lightened; the Sun shone through, pale and milky, and they emerged on one side of the storm. They could see it still looming high behind them, but ahead they looked out on a dazzling world of brilliant blue sky and mountainous white clouds that looked as soft and inviting as enormous piles of down. They were at 10,000 feet, Glaisher announced as he checked his instruments. Miraculously, none of them had been lost nor had their equipment been damaged in the storm. The temperature hovered around the freezing point and the men buttoned up their coats to protect themselves from the cold wind. Coxwell emptied three bags of sand over the rim of the basket and the balloon rose swiftly in response.

Glaisher began taking his measurements as calmly and methodically as if he had been snug and warm in his laboratory. When they reached 15,000 feet, the water in the wet-bulb thermometer began to freeze, making the instrument useless. Its last measurement was 18 degrees Fahrenheit. Glaisher released one of the pigeons. It extended its wings, then fluttered awkwardly and somewhat out of control toward the earth.

At an altitude of 20,000 feet a second pigeon was released. It fell a considerable distance before it began to fly very slowly downward in a spiral pattern.

At 25,000 feet, the temperature had dropped far below zero. Both Coxwell and Glaisher had difficulty performing the simplest tasks. It was hard to focus on the instruments and to record the numbers, and it was an almost impossible task to pour out another bag of sand. Coxwell opened the pigeons' cage to let out the last two birds, but they were huddled together at the back of the cage. Finally he shoved them out. One fell straight downward, never opening its wings at all. The other fluttered frantically, managing at last to fly upward toward the top of the balloon, but it was never seen again.

The balloon was still rising. It had reached 28,000 feet when Glaisher noticed that the cord controlling the release valve was out of reach, entangled in the rigging. There was no way to let the gas out. They could not bring the balloon back down to Earth, at least not until the Sun went down, permitting the gas to cool and shrink. But sunset was many hours away.

Both Glaisher and Coxwell stood up, studying the problem. The basket was hung from a large iron hoop, and the shroud ropes connected the hoop to the balloon. By climbing onto this hoop, Coxwell thought, he could just reach the tangled valve cord. He must try, although every step, every movement was now incredibly difficult. He hoisted himself first onto the edge of the basket, balancing there 28,000 feet above the Earth. Then with the help of the shroud ropes, he pulled himself slowly upward until he was able to hook one leg over the iron hoop. By careful maneuvering, he pulled his other leg across and rested, grasping the hoop tightly to steady himself. It was bitterly cold; the neck of the balloon was white with frost. Coxwell saw that the valve cord was now within reach. He tried to unclench his hands to grasp it but he could not free them. They were frozen fast to the iron hoop.

Watching Coxwell, Glaisher did not understand what had gone wrong. He had trouble seeing clearly; there seemed to be a fog before his eyes. "I dimly saw Mr. Coxwell," he recalled later, "and endeavoured to speak, but could not. In an instant, intense darkness overcame me, so that the optic nerve lost power suddenly, but I was still conscious, with as active a brain as at the present moment whilst writing this. I thought I had been seized with asphysia and believed I should experience nothing more, as death would come unless we speedily descended: other thoughts were entering my mind when suddenly I became unconscious as on going to sleep."

Coxwell called out to Glaisher but got no answer. He tried desperately to drag his hands loose from the hoop but he could not free them. They had turned dark—almost black—with cold. Coxwell felt no pain, but an overwhelming sleepiness. Just over his head the valve line swung in a slow circle. With an enormous effort of will, he lunged and tried to catch it in his teeth. Twice it eluded him. Then, finally, he caught the line in his mouth. He wrenched it so hard that he broke a tooth, but the valve did not open. Still holding the cord in his teeth, Coxwell used his tongue to push it back toward the side of his mouth where he could grasp it with his molars. Then he pulled again. This time the line cut deeply into his cheek. His mouth filled with blood and suddenly nausea overcame him. He vomited, losing the valve line and all the contents of his stomach into space. Finally, the wave of sickness passed. He seized the valve cord again and, clamping his teeth down hard on it and gathering his last ounce of strength, threw himself backward. The valve broke open suddenly and his hands tore free. Somersaulting through space, he landed in the basket.

The line was free now, hanging so low that Coxwell could reach it easily. He caught it again in his mouth and, pulling gently, he heard the soft whistle of escaping gas. The balloon slowly began its descent.

It was many minutes before Glaisher, paralyzed by lack of

oxygen, began to recover his sight. He could hear Coxwell urging him, "Do try; now do," but he could not speak or move. Then, very gradually, as the descent continued, all his faculties returned.

— • —

Both aeronauts recovered from their extraordinary trip, which had taken them to the very limits of the space in which human beings can survive unprotected. Before Glaisher passed out he had taken a reading of 30,000 feet, but had been unable to record the number. This altitude is several times higher than that which is considered safe today.

Millions of years of evolution have adapted man to exist and function in the presence of atmospheric pressure and oxygen concentrations at ground level. At 33,000 feet, the air contains only about one-third as much oxygen, and the pressure is also approximately one-third as great. At 55,000 feet, an unprotected man becomes unconscious in a few minutes, and longer exposure leads to brain damage and death. Today pilots are warned not to fly above 10,000 feet for long in an unpressurized plane without oxygen.

Nature has adapted people who live in very high mountain environments, such as the Indians on the Altiplano in Peru (about 12,000 feet), to the thin air by increasing their lung capacity and the number of red corpuscles in their blood. Thus the more rarefied supply of oxygen is used more efficiently.

During the last decades of the nineteenth century and the first half of the twentieth many of the problems associated with high-altitude flight and balloon exploration were solved. Pressurized gondolas and special space suits were invented, as were methods of carrying supplies of oxygen aloft. Helium was discovered—an element that is lighter than air and inert, so it does not have the explosive property that makes hydrogen hazardous. Very strong, thin plastics made better envelopes for the balloons. More sophisticated instruments and communications increased the effectiveness of exploratory flights. By the middle of the twentieth century, aeronauts were able to reach 100,000 feet in comfort and safety. This proved to be the practical limit of lighter-than-air exploration because the balloon and its occupants weighed as much as the air they displaced at that altitude.

Lieutenant David G. Simons described how it felt when the helium aerostat he was riding in, as part of the scientific project called Man High, reached this altitude. His craft had been rising rapidly and steadily as if "on an endless elevator." Then the rate of ascent began to drop off sharply. The balloon was almost full, a swollen ball 200 feet wide. "The altimeter faltered, then stopped. I noted the time . . . it had taken me two hours and eighteen minutes to bump against the earth's ceiling. Slowly the needle dropped, then rose again. The balloon was gently bouncing, like a basketball being dribbled in slow motion in an upside-down world. It had ascended as high as it would go."

Simons was deeply impressed by the beauty of the view from space. The air was clear, with a transparency unknown on Earth. The sky was a dark, velvety violet, and the horizon a brilliant, iridescent blue. "I saw something I did not believe at first," he said. "Well above the haze layer of the earth's atmosphere were additional faint thin bands of blue, sharply etched against the dark sky. They hovered over the earth like a succession of halos."

It is not immediately obvious why the Earth's atmosphere is layered. Molecules of gas all move very rapidly and tend to even out any differences in temperature and density; so, common sense would tell us that there should be no sharp transitions or layers. We would expect the Earth's atmosphere and all the influences caused by the planet's presence to recede uniformly and merge imperceptibly with outer space. This common-sense assumption was the basis of the accepted scientific model describing the distribution of the gases around the Earth less than a century ago. It was not until men actually penetrated into this medium with their measuring devices that the complex structure of the Earth's atmosphere was recognized.

The temperature readings proved to be the most surprising. With every 1,000-foot rise in altitude they fell an average of about 3.6 degrees Fahrenheit until readings as low as −80 degrees Fahrenheit were reached. But then suddenly the temperature stopped falling; it leveled out and began to rise slowly. The place where this temperature difference occurred—the tropopause—acts as a ceiling, turning back the rising currents of air and confining 85 percent of the atmosphere in this lowest layer, close to the Earth's surface.

The vertical movement of the colder, heavier air in the upper troposphere is inhibited by the presence of hotter, lighter air just above it. This is called a temperature inversion because hot air lying on top of cold is the inverse of the usual situation, where the temperature of the atmosphere decreases with height for several miles above the Earth.

Local temperature inversions do occur at ground level, too. At night the Earth cools off more rapidly than the atmosphere, and the air in contact with the ground is also cooled. This heavier layer remains near the surface, creating a blanket of air that does not dissipate until it has been warmed by the Sun. This typical nighttime inversion occurs on more than half of the nights throughout most of the United States. It is a phenomenon that contributes seriously to air pollution problems because, in this stagnant surface layer, chemicals that have been poured into the atmosphere remain close to the ground all night. In some situations, episodes persist and build up for several days. These conditions have been responsible for some of our worst pollution

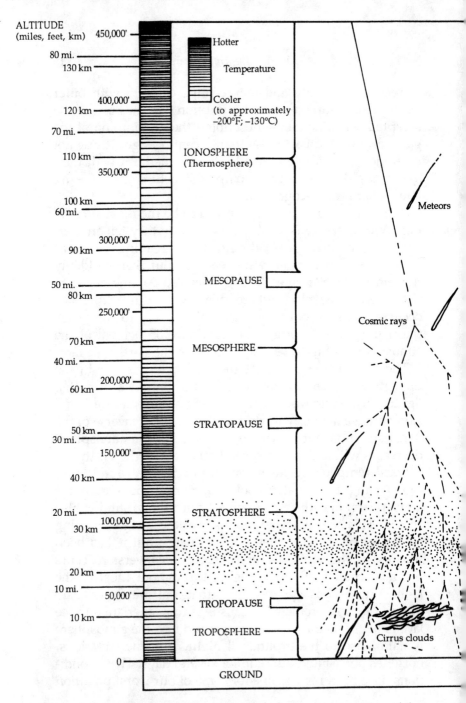

ALTITUDE
(miles, feet, km)

Hotter

Temperature

Cooler
(to approximately
–200°F; –130°C)

IONOSPHERE
(Thermosphere)

Meteors

MESOPAUSE

Cosmic rays

MESOSPHERE

STRATOPAUSE

STRATOSPHERE

TROPOPAUSE

Cirrus clouds

TROPOSPHERE

GROUND

Figure 1. This diagram shows an approximate cross-section of the lowest eighty-five miles of the Earth's atmosphere in the midlatitudes. There are seasonal variations.

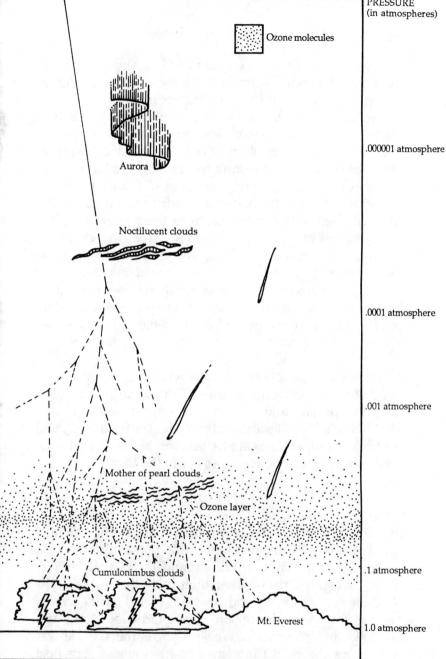

PRESSURE
(in atmospheres)

Ozone molecules

Aurora

.000001 atmosphere

Noctilucent clouds

.0001 atmosphere

.001 atmosphere

Mother of pearl clouds

Ozone layer

.1 atmosphere

Cumulonimbus clouds

Mt. Everest
1.0 atmosphere

17

problems in cities such as London and Los Angeles. In 1948, there was a famous episode in Donora, Pennsylvania, in which 40 percent of the population became ill.

The great temperature inversion that surrounds our planet serves many useful functions that benefit life on Earth. It helps to control the circulation of our atmosphere, distributing heat and moisture more uniformly. But because of the existence of this inversion layer, chemicals broadcast in the atmosphere are not dissipated; they are just redistributed. Before the Industrial Age the amount of these additives was very small compared with the great volume of air in the troposphere, and the effects were too minor to cause concern. However, as human populations have grown and industry has poured ever larger quantities of polluting substances into the atmosphere, these pollutants have become a threat to health and even to the Earth's weather system. They are truly seeds of destruction that we have sown upon the wind.

Above the troposphere are layers where tenuous gases are laced with energy from the Sun. With increasing altitudes, temperature rises and falls and rises again (see Figure 1). The coldest regions lie about fifty miles over the tropics, and the hottest temperatures are found over the frozen Antarctic and Arctic lands at ninety miles above the Earth.

— • —

But David Simons, looking out upon the "Earth's ceiling," could see only the faintest hint of the powerful forces that create this many-layered halo around the planet. He was above the realm of the weather; high above the mists and the rain, the snow showers and the thin, curled cirrus clouds. He was far above the region where tornadoes form and rainbows arch their delicate colors across the sky. He was even above the giant thunderheads from whose tops float streams of ice crystals in shining anvils. At this point he had reached the highest altitude that could be achieved by balloon flight. The exploration of the atmosphere seemed to have reached a natural limit.

However, just two months later a rocket rose from the grassy steppes of Russia, northeast of the Aral Sea. It carried Sputnik 1 into an elliptical orbit one thousand miles above the Earth's surface. This first man-made star, which astronomers christened 1957 Alpha, carried instruments that measured the Earth's magnetic field and the Sun's energy. Two radio transmitters relayed the information back to Earth. Since that day when the Space Age dawned, knowledge about our atmosphere has burgeoned.

It happens so often in science that, when a final barrier seems to have been reached, another way opens up, leading to frontiers far beyond.

2

— • —

The Realm
of the Weather

Soaring on the wind. (Photograph by Georges Uveges.)

Like a rainbow translated into delicate shades of blue, a succession of halos extends hundreds of miles above our heads. But the first halo—the one in which we live and breathe—is the shallowest of all. It reaches only five to ten miles above sea level. In some ways this is a comforting thought; there is something quite frightening about unbounded immensities. But this slender layer of air is very vulnerable because it is shallow and contained. It holds the breath of all living things and the effluent of all the activities of man. As Lewis Thomas expressed it:

> The word is out that the sky is not limitless; it is finite. It is, in truth, only a kind of local roof, a membrane under which we live, luminous but confusingly refractile when suffused with sunlight; we can sense its concave surface a few miles over our heads. We know that it is tough and thick enough so that when hard objects strike it from the outside they burst into flames. The color photographs of the earth are more amazing than anything outside; we live inside a blue chamber, a bubble of air blown by ourselves.

It would be a mistake, however, to leave the impression that the tropopause acts as a perfect seal between the troposphere and the stratosphere. There are holes in our ceiling, certain breaks that occur at the places where major air masses converge. These holes allow some interchange between the two layers.

The shape of the troposphere changes from month to month and from the equator to the poles. It rises to about ten or eleven miles high at the equator, and decreases to four or five miles at the poles. In the midlatitudes it is typically six or seven miles above the Earth. The higher numbers are characteristic of summer months; the lower ones occur when the air cools off in the wintertime. These variations are easily explained as the effect of heating by the Sun, causing the bubble to expand like a hot-air balloon.

Temperature differences are the main forces moving the winds and weather systems on Earth. The equatorial zone receives the maximum amount of solar energy because the incoming rays here are more nearly perpendicular to the planet's surface. At midlatitudes, the rays make a longer transit through the atmosphere and strike the ground at a glancing angle; at the poles, they are almost parallel to the surface and have little effect. So the spherical shape of the planet results in an uneven distribution of solar energy.

The long, warm days of the tropics are punctuated by afternoon thunderstorms and by squalls, which occur most frequently at the times of the equinoxes, when the Sun shines directly down on the equator. The rising column of warm air is loaded with moisture that condenses out into clouds, and in this process heat is given off, warming the air again. This cycle breeds thunderstorms, which act like enormous pumps sending pulses of energy high into the troposphere. It carries warmth from the Sun and moisture from the tropical seas to altitudes of about ten miles. Here at the tropopause, the updraft ends; the air turns and moves toward the poles.

If the Earth did not rotate on its axis, the direction of flow would be directly north and south, but the turning of the Earth imparts a curved motion to these winds. To understand why this happens, imagine that you are standing on the equator, and that you have just shot an arrow toward the North Pole. Although the arrow is traveling in a straight line in space, to an observer on the surface of the Earth, the path of the arrow would appear to be curved because it is moving east more rapidly than the ground beneath it. All objects on the equator, including you and the arrow (before it was shot), are moving eastward at a faster speed than objects farther north or south. In Washington, D.C., for example, the easterly component of motion is less than at the equator, and in Montreal it is smaller still. But the arrow retains the easterly component of motion it possessed when it left your hand, and by the time it reaches Washington, D.C., it is moving east faster than the terrain below it. Therefore, in relation to the ground, it curves to the right of its path as it travels north. If the arrow is aimed at the South Pole, it will also curve toward the east but, this time, to the left of its path. In the Northern Hemisphere all moving objects, including the winds, are deflected to the right of their trajectory by the Earth's motion and, in the Southern Hemisphere, to the left.

Cooled by their passage toward the poles, the tropical air masses begin slowly to sink again. In the meantime, the motion of the Earth has deflected them, so that by the time they reach the twenty-fifth or thirty-fifth degree of latitude (for example, between San Antonio, Texas and Phoenix, Arizona) the air in these bands is traveling almost due east. As it sinks it causes high-pressure areas, forcing the air in the lower atmosphere to move out. Nudged by this high pressure and "attracted" by low pressure near the girdle of the Earth, surface winds flow toward the equator, and as they move they are deflected to the right of their path in the Northern Hemisphere and to the left in the Southern. Thus the easterly trade winds are created, completing the circulation of the tropical air.

The latitudes around 25 to 35 degrees, where the downward air movement is dominant, have almost continual high-pressure conditions. These are fair-weather zones with clear skies and exceedingly low humidity because most of the moisture is condensed out when the air rises to very high altitudes in the equatorial zones. As the air sinks again, it is warmed by compression, resulting in further decrease of relative humidity. The greatest deserts of the world fall into these bands: the Sahara, the Arabian, the Kalahari, the southwestern desert of the United States, and the Great Victoria Desert in Australia.

The areas of persistent high pressure are also especially likely to suffer from man-made pollution. Because upward motion of air is suppressed, smoke and fumes have a longer residence time at ground level. Los Angeles (at 34 degrees north latitude) is an outstanding modern example of a city built in a region naturally blessed with a wonderful climate but which has been marred by pollution held close to the Earth by persistent high pressure.

While the flow of air toward the equator creates the trade winds, a flow also occurs in the opposite direction, carrying warm, subtropical air farther north, where it encounters cold air masses. The bands of latitudes between 35 and 60 degrees are known as the temperate zones—an ironic title, because these are the regions of the most variable and unpredictable weather conditions with great extremes of both wind and temperature. Within a few hours, or even within a few minutes, the thermometer can plummet as much as 50 degrees. In North Dakota, on the plains of western Canada and the steppes of Russia, the temperatures normally swing between 110 degrees Fahrenheit in summer and −50 degrees Fahrenheit in winter, with winds up to sixty miles an hour. The great variability of the climate in these latitudes is caused by the alternating domination of polar and subtropical air masses.

In the polar regions, easterly winds usually prevail at low altitudes. The air, which is chilled over the frozen top and bottom of the planet, sinks, causing high pressure and forcing surface winds to flow down and outward toward the midlatitudes. Here, the polar air encounters the flow of humid subtropical air. A broad band of turbulence is produced in each hemisphere, with warm and cold fronts, moving on westerly wind currents around the globe. In the lower atmosphere the turbulence produces large swirls and depressions.

Controlled by the global wind patterns and the rotation of the Earth, the low-pressure whirls rotate counterclockwise in the Northern Hemisphere and are known as cyclones (this is a general term not to be confused with hurricanes, which are called tropical cyclones). In the Southern Hemisphere they rotate clockwise. High-pressure areas (anticyclones) are just the opposite, having winds that travel clockwise in the Northern Hemisphere and counterclockwise in the Southern. These are the typical high- and low-pressure systems that troop in an endless procession around the midlatitudes, producing the wild, winter blizzards, the humid heat of summer, and the destructive ice storms of the "temperate" zones. But they also bring abundant precipitation throughout the year, making these areas especially suitable for farming. Here lie the great breadbaskets of the Earth, such as the midwestern United States and the Russian Ukraine.

The path of these weather systems varies greatly from winter to summer and even from week to week in a way that has been difficult to predict. But an important clue to understanding and anticipating their path was discovered during World War II. Sometime about 1942, the Japanese noticed a remarkably swift current of air that blows eastward at great heights over their islands toward the west coast of America. This "jet stream" was probably encountered by their bomber and fighter pilots heading for Hawaii and Midway Island. But it was a carefully guarded secret because of its military importance. This wonderful natural superhighway in the sky could reduce the flying time from Japan to the United States by six or seven hours and save thousands of gallons of fuel for one plane alone. Better still, it could carry a balloon offensive to the American mainland without using any fuel at all and without endangering the lives of any Japanese pilots.

With this knowledge, during the last year of the war the Japanese carried out an imaginative scheme. They constructed special balloons of long-fiber rice paper, filled them with hydrogen, and equipped them with an ingenious device which regulated their altitude so that a balloon launched at a height of 30,000 feet over Japan might stay within the jet stream until it reached the coast of the United States. When the altitude, measured by a barometer, dropped below 29,000 feet, an automatic mechanism released one of twenty small sandbags. If the balloon rose above 36,000 feet, a valve triggered by the barometer released some of the hydrogen gas so the balloon would sink again. Since the jet stream was always found within this band of altitudes, the balloon should remain in the stream and would ride its current for 5,000 miles to America. When it reached the coast of the United States, it was programmed to start dropping its cargo of incendiary and antipersonnel bombs. As the last bomb was dropped, a block of picric acid was designed to explode, destroying all the instruments, and a pouch of magnesium flashbulb powder would flare up, turning the balloon into a ball of fire.

The Japanese had invented a remarkably clever weapon, but they made one mistake. The electronic equipment was powered with a wet-cell battery which froze in the subzero temperatures of the jet stream, and, as a result, only about one-tenth of the 9,000 balloons released in Japan ever reached the United States. Even with this serious defect, the balloon offensive was successful enough to give the U.S. military establishment a very bad scare.

Enough of the balloons reached land to cause great concern and some serious consequences. A balloon landed on a beach in Oregon where it was discovered by a group of children enjoying a Sunday-school picnic. The large, crumpled object lying on the sand attracted their attention. Several of the children poked at it and started to turn over one corner. Suddenly the object burst into flames and exploded in their faces, killing five children and the pastor's wife who had been in charge of the ill-fated outing.

Bomb blasts were heard in remote places as widely separated as Thermopolis, Wyoming and Ventura, California. In downtown Los Angeles tattered bits of rice-fabric paper were picked up on the streets. Reports from Mexico, Canada, and Alaska mentioned seeing as many as seventeen balloons in one day, many of which had dropped incendiary bombs. Throughout the western states, forest fires suddenly became so numerous that the Army was called in to help fight them. Some members of the High Command feared that these aircraft would soon be used to carry bacteriological warfare to the United States.

American pilots flying over the Pacific and on bombing missions over Japan reported seeing large silver spheres floating high over the cities. They thought these might be some new kind of antiaircraft device, but the spheres were flying at heights greater than the altitudes normally used by bombers, and many of them appeared to be drifting out to sea. The military establishment in the United States did not believe that the balloons found in this country just a little while later could be the same ones that were seen floating over Japan. It seemed impossible that balloons could have traveled such immense distances in such a short time.

Very few of these strange and alarming reports reached the American public because a voluntary censorship had been agreed upon by the press and radio in the United States and Canada. This action was largely responsible for defeating the balloon offensive. Only two or three balloon sightings were reported by the media. Consequently, the Japanese General Staff believed that of the 9,000 launched in Japan, only these few had reached America. After six months, they discontinued the operation.

In the meantime, the jet stream itself had been discovered by American pilots. In November 1944, the first high-altitude bombing mission over Japan was undertaken. As 111 B-29 planes, flying between 27,000 and 30,000 feet,

turned east near Tokyo, they were suddenly swept forward
by a 150-mile-per-hour wind. With ground speeds of 445
miles per hour, the bombardiers were not able to compensate for wind drift. Only sixteen of the hundreds of bombs
they dropped hit the industrial target. The rest fell on the
city or in the sea.

On other occasions, Air Force pilots flying west encountered such high winds that their planes seemed to stand still
in the air. Islands that should have been passed long ago
remained stationary below them as though the whole scene
had frozen. Although the engines were racing at full throt-
tle, the planes were not going anywhere, and in some cases
the pilots had to turn back before their load of fuel was
exhausted.

After the war, scientists from all over the world followed
up these discoveries with a cooperative international effort
to reveal the secrets of these remarkable currents of air and
to find ways of using them constructively. A network of
instrumented balloon flights and specially equipped aircraft
probed the upper troposphere and lower stratosphere, plot-
ting the position and speed of the winds. The nature of the
jet streams is now quite well understood, and their position
is followed regularly from day to day.

It turns out that there are four principal jet streams. They
are located near the tropopause, where air masses of quite
different temperatures meet. These are the places, you remember, where breaks occur in the tropopause—the place
where the polar air meets the subtropical and the region
where tropical air meets that of the temperate zones. Jet
streams form at or just below these breaks in the tropopause.

The jet that flows in the break where the polar air meets
the subtropical is known as the *polar-front jet,* although it
usually occurs in the midlatitudes. This is the stream that
carried the fire balloons across the Pacific. It is the lowest of
the jet streams because the level of the tropopause at this
point is only six or seven miles high. It is not surprising,

therefore, that it was the first jet stream to be discovered. The winds of the polar-front jet travel from west to east around the globe, following a sinuous wave pattern because the meeting place of the air masses is influenced by regional and seasonal temperature differences. In summer, when warm air predominates in North America, for example, the polar-front jet moves to higher latitudes, crossing North America mostly in Canada. In winter, it moves farther south, traversing the United States. Its movement with the seasons corresponds with the seasonal shift of the band of turbulent low-atmosphere westerlies. In fact, these two phenomena are now believed to be different aspects of one system.

— · —

The cyclones and anticyclones, which flow from west to east, are triggered by changes in the velocity of the jet stream. Pulsations in wind speed within the core of the jet move slowly along the stream's path, migrating eastward at an average speed of twenty-two miles per hour. Increases in speed cause air to flow out of the stream, just as a brook might overflow as it picks up speed going downhill. Then pressure decreases in the vertical column of air below that point; the barometer drops and a cyclone develops at ground level. Conversely, when the jet stream slows down, air flows back into it, causing higher pressure under that point on the surface of the Earth. This high pressure creates an anticyclone. And the whole system of highs and lows moves slowly eastward.

The subtropical jet stream flows in the region between the midlatitudes and the tropics, but it occurs only in the winter when the most extreme temperature differences exist on the planet. It moves at higher altitudes than the polar-front jet and has less influence on ground-level conditions.

These two wind systems are duplicated in the Southern

Hemisphere, making up the four most important jet streams. They all flow from west to east, through a space several hundred miles wide but only two or three miles deep, and wind speeds at their cores as high as 300 miles an hour have been measured.

During the summer months in the Northern Hemisphere, when the subtropical jet disappears, another type of jet wind forms. It flows from east to west, at very high altitudes—in the lower stratosphere. This jet brings the characteristic monsoon rains to southern Asia and Africa. The exact timing of the change from the domination of the westerly subtropical jet to the easterly jet is very critical to the agriculture in these lands of marginal rainfall. A delay in the reversal of the wind pattern can cause crop failures, resulting in the starvation of millions of people. A southerly shift in the path of the easterly jet is even more serious because it moves the belt of monsoon rains out of northern India and the sub-Sahara. This shift can be caused by a very small increase in the temperature contrast between the Arctic and the tropics.

Knowledge about these remarkable rivers of air is an invaluable aid in making more reliable weather predictions and in monitoring the long-term changes that are occurring in the planet's weather. The position and speed of the streams (especially the polar-front jet) help to foretell the location and the violence of the storms that will occur on Earth. A plot of their movement with the seasons provides an excellent way of taking the world's temperature because the position of the jet streams is so closely related to alterations in heat flow between the tropics and the pole.

The moving pattern of the jet streams and their associated bands of high- and low-pressure systems is superimposed upon the more permanent system of high- and low-pressure regions created by the geography of the planet: mountain ranges and deserts, regions of persistent cold or hot weather.

To understand the interaction of all these factors as well as the day by day and seasonal temperature changes is a constant challenge to meteorologists.

At noon and midnight Greenwich Mean Time, approximately seven hundred weather stations around the world launch radiosonde balloons containing miniature weather stations that transmit information about the upper air as they rise toward the top of the troposphere. Carrying radar reflectors, they can be tracked even through clouds to yield measurements of temperature and wind speed and direction. All of these data are relayed by the Global Telecommunications Systems around the world. Together with the information from satellites and ground-level observations from about nine thousand other weather stations, these facts are fed into giant computers. This global weather watch provides an impressive amount of information almost continuously, thereby enabling meteorologists to plot the wind-flow patterns of the Earth. They can usually pinpoint the position and estimate the strength of the jet winds.

It is hard to believe that just half a century ago men did not dream that rivers of air were passing overhead at such fantastic speeds. To the unaided eye they give little or no sign of their presence—occasionally leaving just a thin trail of cirrus clouds in their wake. But, most of the time, they pass silent and swift as ghosts across the sky.

— • —

From Earth-side, looking up, men have tried for thousands of years to understand the messages written in the clouds. Are they harbingers of destructive storms or will they bring the anxiously awaited rain? All clouds, from the tiniest wisps of fog to the solid layers that blanket thousands of miles, mark places where water vapor has condensed into tiny cloud droplets. Water that is present as a gas in the air is entirely invisible. The separate molecules are so small and widely spaced that light passes by them without being significantly deflected by their presence, and the air that contains them may be perfectly clear and transparent. When the air is warm, more water can exist there in vapor form than at lower temperatures. As the air cools, the water molecules tend to condense back into liquid form, making tiny droplets of water that are still small and light enough to be airborne, suspended by the rapidly dancing molecules of the gases in the atmosphere. Fifty billion of these droplets would not fill a teacup; but they are the substance of all the clouds in the sky, from the filmy curtains of cirrus to the towering cumulonimbus. In the process of condensation heat is given off, warming the surrounding air and conversely, during evaporation when liquid water is converted back into vapor, heat is absorbed from the environment.

It was noticed very early that certain types of cloud formations are usually associated with particular weather changes. The first step in understanding these relationships was the classification of clouds.

The cloud names commonly used today are derived from Latin: *cumulus* means a heap or pile; *stratus,* a layer; *cirrus,* a curl; *nimbus,* rain; and *alto,* raised or elevated. The first three designate the principal cloud types and the last two are used as qualifying terms. Within this small vocabulary the myriad cloud effects are described. *Cumulonimbus* means the great heap of cloud that produces a rainstorm; *altostratus* is a high, uniform layer of cloud; and so on. There are many possible combinations, and fine distinctions exist between the various types.

Clouds are formed when air becomes too cool to contain as water vapor the amount of moisture it is carrying. The temperature at which this occurs varies greatly, depending on a number of factors—the humidity, for example, and the dust load present in the atmosphere. Finely divided particles of foreign matter in the air serve as nuclei around which condensation takes place most efficiently. Very dry, clean air over the desert must be cooled to about 35 degrees Fahrenheit before condensation takes place. Since this temperature drop rarely occurs, the desert skies are almost always clear. On the other hand, extremely humid air may begin to turn into a cloud when the temperature is as high as 80 degrees Fahrenheit. The change of just a degree or two is enough to tip the balance. Under these conditions clouds form and dissipate right before our eyes like phantoms materializing out of thin air and then, without a word of warning, become invisible again.

Cumulus clouds are the ones that form most commonly in this delicately balanced atmosphere. Air carrying moisture from the tropical seas rises when it drifts over land surfaces warmed by sunshine. For this reason, fluffy white cumulus clouds appear most frequently over islands in hot, humid climates and build up throughout the afternoon, only to dissipate as the sun goes down and no longer heats the face of the planet. The South Sea Islanders who ventured thousands of miles in fragile open canoes used these markers in the sky to guide them to new land. New Zealand was known as "the land of the long white cloud" years before it was colonized.

These cumulus clouds, like piles of white froth, are among the most attractive of all formations. When large numbers of them are massed, they shadow each other and create a fantastic landscape with deep purple-shaded valleys, long avenues of luminous pillars, and mountain peaks radiant with sunshine. Sharply delineated flat bottom edges occur at the condensation level, usually from 2,000- to 3,000-feet high, but up to 12,000 feet in high plains and desert areas.

Under every cumulus cloud there is an updraft that feeds warm, moist air into the center of the cloud. Glider pilots have discovered how to take advantage of these natural elevators to gain the altitude needed for extended flight. If conditions are just right and little white puffs can be seen developing in a blue sky, pilots can be heard saying to each other enthusiastically, "The cums are popping!" But these innocent-looking clouds must be treated with caution. As the hours pass and the cloud formations develop, very powerful thermals can sweep the defenseless sailplane straight up to the top of a developing storm.

Such a tragic event occurred during one of the early gliding contests in central Germany. Northeast of Frankfurt, near the border of East and West Germany, lies a range of mountains, which creates wonderful gliding conditions— strong thermals and ridge lift generated by air moving up the mountainsides. On a warm summer day in 1938, a number of happy glider pilots gathered there with their streamlined sailplanes to take part in a contest to achieve the highest altitude. Large cumulus clouds had formed, giving promise of great opportunities for record-breaking flights. In fact, heights of more than 26,000 feet were reached. But five contestants, carried away by the spirit of competition, flew too close to the base of a fully formed cumulonimbus. All five were suddenly sucked into the center of the vortex. The sailplanes were jerked up into a violent squall and driving rain blotted out all visibility. Because the mountains were very close by, the pilots feared being thrown against a cliff. They panicked and jumped from their planes, pulling the release cords of their parachutes.

The consequences were disastrous. Instead of descending gently, the parachutes, driven by the howling force of the wind, carried them ever higher and higher into air so cold that ice formed on their bodies. Hailstones lashed their faces. At heights that probably exceeded forty-five thousand feet, they were encased in frozen water—living icicles tossed in the turbulent air and blinded by lightning. When the storm finally relaxed its grip, four lifeless bodies fell out of the cloud. Only one man reached the ground alive and he was severely injured.

— • —

Stratus clouds are much less dangerous and less attractive visually than the cumulus. Their flat structure is formed when air moves in a very uniform manner, rising slowly to a location where a cooler temperature causes condensation. Stratus clouds are usually found where a warm air mass passes over colder air. The warm air, being lighter, rides up over the cold air and its moisture rapidly condenses out in a broad uniform layer. These clouds generally create a dark-gray pall across the sky and, since they bring hours of rain, they are known as nimbostratus.

Sometimes the clouds comprising the layer are broken up into a mottled structure caused by a wave type of circulation within the shallow cloud layer; the air rises in one place and descends in another, and so on, at regular intervals. This circulation pattern breaks the cloud layer up into cells and produces the typical "mackerel sky," which is an omen of bad weather. Stratus clouds occur at many different altitudes, from 30,000 feet down almost to ground level.

Cirrus clouds, on the other hand, always ride very high in the sky at five to nine miles above the Earth's surface. They reflect the warm colors of the sun immediately before dawn and, after sunset, leave the rest of the world in darkness. These ephemeral cloud formations are composed entirely of ice crystals and are so tenuous a form of matter that they contain only a few ice crystals in each cubic inch. They are too insubstantial to block out the light or to cast any shadows. The beautiful shining filaments and mare's tails that describe curved patterns across the sky are ice cascades seen from underneath as the crystals fall through the thin air of the upper troposphere (see photograph on page 4). Because of their great height they appear to move slowly, even though they may be traveling 100 miles an hour or more. Cirrus clouds indicate the presence high aloft of air whose moisture content (and therefore whose origin) is different from the air immediately beneath the cloud. They form in regions where there are updrafts that carry water vapor very high and are usually the first sign of an approaching low-pressure system, perhaps 500 or 600 miles in advance of the front.

The first astronauts looking back on our planet from outer space were all impressed by the unexpectedly large amount of cloud cover, moving in a restless swirl across oceans and continents. It veils roughly half of the Earth's surface.

— • —

One very important question about the Earth's cloud cover has not been definitively answered. Does this great mass of cloud cool or warm the Earth? We have all noticed, of course, that on cold nights the presence of cloud cover moderates the temperature drop and often prevents frost. Conversely, a cloud passing overhead and blocking out the Sun causes a sudden chill in the air. All clouds act to some extent as reflectors, turning the incoming sunlight back into space, while simultaneously warming the Earth by preventing heat from escaping. Different clouds combine these characteristics in different degrees. Dense, low-lying stratus clouds have a strong cooling effect; tenuous high-riding cirrus clouds are believed to warm the Earth because they are semitransparent to incoming radiation but block some of the heat radiation emitted by the Earth.

Another factor, which must be taken into consideration in estimating the total effect of cloud cover, is the reflectivity of the Earth's surface. This varies greatly from snow-covered land to dark rain forests to the surfaces of the sea. Stratus clouds covering a snowy landscape, for example, would have a smaller net effect on the Earth's temperature than the same cloud formation over a tropical forest. For decades scientists have been studying these relationships, but the problem is very complicated and no firm conclusions have been reached.

Results of a very interesting study were announced just recently by a team of scientists at the University of Chicago. Fourteen years ago they began working with the National Aeronautics and Space Administration on a project to measure, by satellite, the radiation received by the Earth as well as that which it loses to space. Since 1984, sophisticated instruments designed by this team have been sent aloft in satellites to collect data. A mass of information has been processed, and a computer analysis generated, simulating by time sequence the net heating and cooling effects of cloud systems for the whole planet. Results suggest that the clouds surrounding the Earth today have a surprisingly large net cooling effect. Without this cloud cover, the Earth might be warmer by 20 to 40 degrees Fahrenheit.

However, the scientists conducting this study are cautious in their statements, because they believe that a change in the nature and location of the clouds in the atmosphere could alter the way in which they affect the temperature of the planet. We will discuss these results in more detail when we consider the greenhouse effect.

— · —

There is some historical evidence supporting the view that cloud formations cool the Earth. Major volcanic eruptions eject quantities of aerosols and particulates with great force into the atmosphere. These are carried very high—even penetrating into the stratosphere—and they act as triggers, causing the condensation of water vapor into clouds. Large eruptions have been followed by a year or two of cooler weather over large areas of the world. In 1815, the Tambora eruption in Indonesia threw a veil of dust into the stratosphere that made 1816 famous as "the year without a summer." New England and eastern Canada got widespread snow in early June, and frosts each summer month. In England, certain regions had rain on all but three or four days from May to October.

The eruption of Krakatoa in the Sunda Strait near Java in 1883 threw approximately thirteen cubic miles of finely divided material into the atmosphere. Some of it was lofted into the stratosphere, and it was carried for such immense distances that red sunrises and sunsets were seen on the other side of the Earth for many months afterward. These colored skies were a sign that sunlight was being scattered by the volcanic dust. The eruption was followed for several years by poor growing seasons in many lands. A sharp freeze occurred in California in 1884, as evidenced by studies of tree rings.

In 1963, after the eruption of Mount Agung on Bali, observations at Mauna Loa Observatory in Hawaii showed a decrease of about one-half of a percent in the solar radiation reaching the troposphere. In recent years, very accurate tracking techniques have been used to follow the spread of volcanic clouds and measure their altitude.

Although nobody understands why, waves of volcanic activity have appeared throughout geologic history. Some climatologists believe that episodes of strong volcanic activity both generated and helped to sustain the great ice ages. Studies of glacial cores show that, at several times in the Earth's history, a high level of volcanic activity occurred simultaneously with minimum temperatures, and the amount of volcanism associated with the last ice ages was greater than it had been during the previous 20 million years.

This evidence suggests that increased cloud cover would cause a net cooling of our planet. But on the other hand, we have before us the example of our sister planet Venus, whose surface we know is hot enough to melt lead, and its atmosphere is so loaded with clouds that they completely obscure the surface. As the space probe Magellan speeds on its way to study Venus in more detail, we cannot help wondering if the condition of this planet should be an object lesson for residents of Earth.

— • —

While climatologists debate this issue, we might turn our attention to the conditions we know would be changed should a denser canopy of clouds cover the Earth. Suppose we lived on a planet completely veiled by clouds. Imagine what life would be like if we could never see the stars at night or watch the Moon rise, drawing a silver wake across the ocean's face. We would never enjoy the rose-colored sky that heralds each day, or watch the Sun sink behind the curved edge of the Earth and send long rays like searchlights high into the sky to illuminate the iridescent mare's tails and the mother-of-pearl clouds. Wrapped in a dense veil, Earth would be a much less beautiful place to live.

The amount of cloud vapor that floats in our atmosphere today provides a most felicitous combination of enfolding softness and translucence that opens up vistas of worlds far beyond our own. It holds and refracts the sunlight, adding glowing color where there would otherwise be blinding white light. It makes our gentle transition hours of dawn and dusk, our rainbows, and the infinite variety of our days.

But most important of all, a heavier cloud cover would affect all life on Earth because life feeds upon the energy of the Sun. If the amount reaching the Earth were significantly diminished, many forms of living things would not survive.

Our great forests would die back, our fields of grain would yield sparse crops, and coral reefs would look like piles of bleached bones in the sea.

One of the most favored explanations for the Great Extinction that occurred about sixty million years ago is the effect of a very large meteorite striking the Earth, throwing up vast quantities of dust, which triggered the formation of a heavy cloud cover and reduced the flow of sunlight for several years. This change was sufficient to wipe out the dinosaurs, the large marine reptiles, the ammonites, and most of the marine plankton.

We cannot afford to ignore these warnings. We must guard this halo of radiant air in which we live—what Shakespeare called "this most excellent canopy . . . this brave, o'erhanging firmament, this majestical roof fretted with golden fire."

3

— · —

The Realm of Living Things

Magnolia blossom. (Photograph by Imogen Cunningham.)

A visit to the Antarctic continent is like a trip back in time to the childhood of the Earth before the first green things were born. The Antarctic is all black and white and blue, with startlingly beautiful landscapes—ebony cliffs that are hung with cascading glaciers like frozen waterfalls, and somber fields of jagged, black lava. The icebergs floating in the dark seas are white, variegated with many shades of blue, from deep, cobalt-blue crevasses to aquamarine shadows. The air is so clean; no haze obscures the sharp outlines of the scene. The patterns are clear and shining like a newly minted coin.

Even the life forms that inhabit this unspoiled part of the Earth are black and white or gray: the penguins; the seals; the fragile, white snow petrels; and the albatross that serenely describes figure-eights in the pale, luminous sky of the midnight Sun. The simplicity of the living community is reminiscent of the very early years of our planet when only a few species existed and these proliferated exuberantly in the sheltering environment of the sea.

After visiting this scene for several weeks, I was airlifted back to the familiar world and set down in a green valley flooded with sunshine. Although still far from home, I felt the same rush of emotion that I feel when I come home after a long absence. All around me stretched green fields, a verdant patchwork quilt of many shades, ringed with darker, deeply wooded green hills. Green—green was everywhere. It is the most vibrant, the most vital of all colors, especially the achingly intense yellow-green of new life in the spring: wheat fields reborn after the snows are gone, the fringed blossoms of the maple trees, and the long flowing tendrils of the weeping willow trees. When the Sun shines on palm fronds or through the translucent leaves of a fern

45

tree, one can almost see the solar energy being transmuted into life. And, in fact, green chlorophyll is the magic molecule that makes all advanced life on Earth possible.

— • —

Before the appearance of chlorophyll, the world must have been colored as the polar regions are today—almost all black and gray and blue. There were no flowers, no brightly plumed birds, no richly variegated mantle of grasses to cover the naked rocks of the continents. The wind and the rain, the pounding of waves, and the rumble of thunder were the only sounds that broke the stillness of that pallid scene where no life stirred, no spontaneous movement ruffled the surface of the sea. The first very simple forms of life that evolved in water must have lived and multiplied far below the surface, hiding in that dim world from the ultraviolet sunlight that beat down pitilessly on the barren land and the sterile layer of surface water. These rays are so energetic that they would have damaged the delicate molecules of living matter and even broken up the organic molecules floating near the surface. The earliest life forms must, therefore, have lived in the half-lighted depths so water could act as a shield against the ultraviolet rays.

It has been estimated that about thirty feet of water was required to provide adequate protection. These tiny self-replicating creatures were unable to utilize the energy of sunlight to maintain their own life and movement. They probably subsisted on chemical energy or possibly on the supply of organic molecules that had accumulated in the sea. As these supplies became depleted, their numbers were limited. Gradually they mutated into more complex forms and then, one day, an organism appeared that contained a new form of molecule—chlorophyll. The first true plant had been created, an organism that could manufacture its own food using sunlight—an ability that gave it an enormous advantage over its neighbors. This critical event in the history of the planet must have taken place at least 3.5 billion years ago. Rocks of that age in the Warrawoona Formation of western Australia carry the imprint of primitive photosynthetic organisms similar to some types of bacteria that exist today.

Throughout the next few billion years, a remarkable interaction took place among the atmosphere, the water, the crust of the Earth, and the burgeoning mass of living matter. In a sense, life began remodeling the environment to suit its own needs. And in the process it altered the other three elements of the *biosphere*—the envelope of air, water, and soil in which life exists.

In the biosphere the most common constituent is oxygen. It accounts for one out of three atoms in every molecule of water in the oceans, rivers, and clouds. It is present in combination with calcium, silicon, and many other elements in the rocks and sand of the Earth's crust. It is one of the fundamental constituents of living matter, comprising about one-fourth of all the atoms in organic molecules. And as free oxygen, it makes up about 21 percent of the total volume of atmosphere today. But when life first began on Earth there was no free oxygen in the air.

Gradually over many millions of years the level of oxygen in the atmosphere built up as one of the products of photosynthesis, and at the same time ozone (the three-atom molecule of oxygen) also began to accumulate. This atom has the special characteristic of absorbing the very energetic component of sunlight and thereby shielding fragile organic molecules, which would be damaged by exposure to these rays. When the concentration of oxygen reached about 1 percent of the present level, perhaps as long as 2 billion years ago, ozone shielded out enough of the ultraviolet to allow organisms to live in the upper sunlit layers of the lakes and seas.

Photosynthesis could take place much more efficiently there and the rate at which free oxygen escaped into the atmosphere increased geometrically.

But free oxygen was both a friend and a foe to early life forms on Earth. It is extremely reactive, combining spontaneously with organic compounds. This characteristic is an advantage in the metabolism of sugar and starch, for example. The reaction with oxygen releases large amounts of energy which can be used by the organism. On the other hand, this reactivity makes oxygen so potentially dangerous to fragile living tissue that all organisms that use it had to develop chemical ways of shielding themselves from its destructive nature. Many of the simple one-celled creatures that existed in the primordial seas would have been killed by direct exposure to an atmosphere containing oxygen. (Louis Pasteur discovered that certain very sensitive microscopic organisms cannot tolerate oxygen concentrations above 1 percent of the present level.)

Some of the primitive organisms continued to exist in old ways that protected them from contact with this dangerous element. They stayed deep under water, buried themselves in mud, or hid themselves under layers of sediment. Their descendants—the anaerobic bacteria, yeasts, and some fungi—are still plentiful today, exploiting this ancient way of life.

Other organisms found a more creative solution. Through the gradual process of mutation and evolution, they developed the protective mechanisms that enabled them to use free oxygen safely, a feat that was achieved with the help of *enzymes*.

Enzyme is the name given to a type of *catalyst* that occurs in living organisms. A catalyst encourages and accelerates one specific chemical reaction without itself being used up in the process. Even though the catalyst may undergo change while the reaction is occurring, it is reformed at the end and is ready to perform the same expediting function for an endless stream of identical reactions, providing a shortcut along which the process can be led again and again. All living things make use of these remarkable molecules to control and direct the complicated reactions that take place within their systems, and many processes that seem to be pure magic turn out to be regulated by catalysts.

Photosynthesis itself is governed by a packet of enzymes working together with molecules of the green pigment chlorophyll. The chlorophyll molecules act like little antennae reaching out to absorb energy from sunlight. A small bundle of this energy is transmitted from one chlorophyll molecule to another until it reaches a location where a reaction controlled by enzymes can occur, and the energy causes the ejection of a high-speed electron from the molecule. This electron can be made to do work and thus light is converted into chemical energy.

In a similar manner, enzymes direct the reactions that occur when molecular oxygen enters a living cell. They encourage the reactions that make oxygen useful to the organism and release its excess energy in ways that are not injurious. All cells of higher organisms contain such enzymes and the evolution of this regulating system was probably an essential step in the development of the more complex forms of life after oxygen became an important component of the atmosphere.

When the concentration of oxygen had risen to near present levels (probably around 500 million years ago, although the exact timing of this event is controversial), ozone provided sufficient protection to allow life to climb out of the sheltering seas and colonize the rocky shores. Ever so slowly, but as irresistibly as a tide, the flood of green vegetation flowed across the land.

For the next 300 million years green was the dominant color that was added by life to the land surfaces of Planet

Earth. Mosses and ferns, pine and spruce forests, giant redwood and sequoia trees dominated the landscape. But grasses did not exist and there wasn't a flower or blooming plant to be found anywhere on Earth, even as late as the early Cretaceous period when dinosaurs roamed the land and pterodactyls soared over the seas on huge leathery wings.

Then, about 100 million years ago, another important innovation occurred. Angiosperms, the true seed-bearing plants, made their appearance and, in a brief instant of geological time, they spread around the planet. Each seed of the angiosperm is a fully equipped embryonic plant, carrying its own store of food and able to travel far from its parent plant. These new life forms had also developed a new method of reproduction, which did not rely on the vagaries of the wind for pollination but, rather, enticed insects to pollinate a plant by employing a display of attractive colors and fragrant messages sent on the breeze. Along with the flowers, honeybees evolved, and ruby-throated hummingbirds, and bright butterflies like blossoms released from their stems and floating free on the wind. Grasses carpeted the bare soil where the forests had not taken hold; water lilies opened their pink-and-white starlike blossoms, and buttercups turned whole hillsides into drifts of gold.

Earth began to look as various and beautiful as it does today. But most important of all, the angiosperms produced grains and fruits that provided the concentrated food needed for the warm-blooded active animals that were beginning to appear on Earth. Birds replaced pterosaurs in the sky. Swift antelopes roamed the lands where grasses grew. And carnivores such as tigers and wolves preyed upon the herbivores, whose meat was now the most concentrated source of food in the world. The early progenitors of man came down from the trees to take advantage of this new food supply, so rich in energy that they needed to spend only a portion of their lives in obtaining nourishment. The time thus released has led to the spectacular achievements of mankind—time to compose beautiful symphonies and poetry, to build the soaring arches of the great cathedrals, and the rockets that reach for the stars.

— • —

The phenomenon of plant growth is so familiar to us that we do not stop to wonder how large, solid objects can materialize out of what seems to be thin air—or at least air supplemented only with a little water, sunshine, and soil that does not appear to be used up. Every year by this process hundreds of billions of tons of starch, sugar, and protein molecules are built up, providing the food for the entire life-support system on Earth. Approximately half of the dry weight of all this vegetation is composed of carbon derived from the carbon dioxide of the atmosphere.

Carbon is the basic building block of all organic molecules. Its atom can link up with other atoms in an amazing variety of ways. In order to synthesize these carbon compounds, energy is required, and the energy that is stored in the compound can be released again by various processes such as the burning and metabolism that take place when oxygen is present. All animals breathe out the carbon dioxide that is left over from the metabolism of their food. Plants also must use some of the food that they themselves synthesize and, in this process, carbon dioxide is respired. Even the Earth itself breathes out carbon dioxide; large amounts are emitted in every volcanic eruption.

The leftover products of digestion and dead organic material of all kinds are reconverted by fermentation and decay into simple chemical compounds. This transformation is achieved by bacteria and fungi with the help of enzymes, and one of the results is the re-formation of carbon dioxide, which returns to the atmosphere.

This natural carbon cycle maintains the very small component of carbon dioxide in the air—approximately one-third of 1 percent, a low concentration that is favorable to animal life. Experiments have shown that concentrations of carbon dioxide over 1 percent begin to be toxic for laboratory animals as well as for human beings. At 5 or 6 percent the rate of respiration increases; at 8 percent vertigo and dizziness occur; 9 percent can be fatal. Plants, on the other hand, can tolerate much higher levels. In fact most plants grow more rapidly and luxuriantly in the presence of increased levels of carbon dioxide, if adequate supplies of other essential ingredients, such as water, are present. This fact suggests that plants may have evolved in an atmosphere that contained more of this gas than our atmosphere does today. There is evidence that the proportion of carbon dioxide has changed, fluctuating throughout geologic time and, until this last century, gradually declining as plant life became an important factor in the biosphere. Since the beginning of the Industrial Age, the proportion has been increasing—a phenomenon we will examine in detail in chapter 5.

There are also regular seasonal fluctuations in the concentration of carbon dioxide, due to the variations in vegetation with the time of year. In the spring, when there is a great burgeoning of growth and the mantle of green spreads rapidly poleward across one entire hemisphere, the consumption of carbon dioxide exceeds the return. From April through September, the atmosphere north of 30 degrees north latitude (for example, New Orleans, Louisiana) loses about 4 billion tons of its carbon dioxide content. But, large as this effect might appear to be, it reduces the carbon dioxide concentration in the air only by about 3 percent.

Occasionally dead, decaying vegetation is trapped under a layer of earth or sediment and is prevented from completing the breakdown portion of the carbon cycle. The trapped organic materials, gradually compacted and heated over millions of years, form coal, oil, and gas deposits, which are storehouses of energy-packed carbon compounds.

— • —

A carbon cycle similar to that on land takes place in the sea. Carbon dioxide is readily soluble in water. The amount dissolved depends on the concentration present in the atmosphere and on the temperature of the water. Small, primitive plants such as algae use the process of photosynthesis to convert the dissolved carbon dioxide into organic molecules, releasing oxygen just as the plants on land do. A portion of this oxygen remains in the sea in a dissolved state. The colder the water, the greater the amounts of gases, such as oxygen and carbon dioxide, it can contain in solution. As the water warms, some of the gases return to the atmosphere. (Note that this is just the opposite of the absorption phenomenon in the air. Water holds more gases in solution when it is cool, while the atmosphere holds more water vapor when it is warm.) The vast numbers of minute photosynthetic organisms in the ocean provide food for the larger forms of marine life. When these organisms die, bacteria and fungi break down the organic matter, converting it to carbon dioxide and water.

Just as on land, a small fraction of the carbon compounds synthesized by living matter in the sea escapes decay and fermentation—tiny particles of lime, shells, and bones remain intact and fall slowly down toward the ocean depths, creating a steady drift of white flakes like a snow flurry in the sea. Gradually, over eons of time, the skeletons of untold numbers of marine organisms form a thick accumulation on the ocean floor. In shallow water they are crushed and compacted by the pounding of the waves and the tides. The lightest fragments are gradually separated out and washed ashore, piling up in the great drifts and dunes that line the coral beaches of the world. If you pick up a handful of this coral sand and study it carefully you will often find tiny shells still intact, some of them no larger than the head of a pin. Each of these miniature works of art once held a living thing. For a brief moment they participated in the transformation of energy into movement and growth, weaving their own individual form out of the basic components of living matter. Thus they entered into the vast cyclic movements that circulate a few basic chemicals from air and water to living tissue and then back again to the wind and the sea.

Hydrogen, carbon, oxygen, and nitrogen—these four elements make up most of the air and almost all of living matter. Although nitrogen is by far the largest component of Earth's atmosphere (78 percent is gaseous nitrogen), the amount of nitrogen in living matter is comparatively small (less than 1 percent). Given these facts, it is an interesting paradox that the food supply of organisms on Earth is more limited by the availability of nitrogen than any other of these nutrients. Most of the great ocean of nitrogen in the air is in a form that cannot be used directly by multicellular plants or animals. Nitrogen gas, which has two atoms in each molecule, is an extremely stable chemical; it does not readily enter into chemical combinations with other elements. Before reactions can occur, the nitrogen molecule must be broken down into single nitrogen atoms, which then do take part in a great variety of reactions, a process known as "fixing" nitrogen.

Scientists theorize that nitrogen was probably available in a more usable form on the primitive Earth. Ammonia (a compound of nitrogen and hydrogen) is thought to have been one of the principal constituents of the primordial atmosphere. This compound can be very easily assimilated by plants and used to build living tissue. But when free oxygen began to appear and reacted with the ammonia, the end products formed were water and gaseous nitrogen, which is such a predominant component of the atmosphere today.

Supplies of fixed nitrogen are now somewhat limited in the biosphere. One of the best natural sources is lightning. In the searing heat of the electrical discharge, the nitrogen molecules are torn apart and the single nitrogen atoms then combine with oxygen to form nitrogen oxides. These in turn react with water to form nitrates. In this way, a thunder and lightning storm delivers a large amount of fixed nitrogen to the air and soil.

However, the natural process that accounts for the greatest share of fixed nitrogen is accomplished in silence and usually in darkness by some of the smallest and most primitive of organisms. Certain types of bacteria live underground in colonies near or actually on the roots of a number of larger plants—clovers, alfalfa, peas and beans, cycads, and ginkgo trees. These colonies of single-celled organisms cooperate with the plants in a symbiotic relationship. Some types are parasitic, depending on the plants for their energy; others obtain their energy directly from sunlight. All of these colonies provide their host plants with nitrogen in a form that can be used to build protein molecules. This trick is accomplished with the aid of an enzyme called *nitrogenase*.

The details of this enzyme-mediated transformation are not completely understood, but the remarkable nature of the process can be appreciated when we consider that the only other important natural process for fixing nitrogen involves the power of a lightning stroke, a phenomenon that heats the air up to 18,000 degrees Fahrenheit in a fraction of a second.

Chemists have attempted to duplicate this enzyme-mediated process, but the most efficient method that has been invented is the use of a catalyst in combination with hundreds of degrees of temperature and enormous pressures. This same transformation is accomplished at ordinary temperatures by some of the smallest living organisms

(25,000 of these bacteria laid side by side would measure scarcely one-half an inch). And they do their work so efficiently that, although it provides the largest source of naturally fixed nitrogen, the total amount of nitrogenase in the world is probably less than ten pounds.

After nitrogen has served its purpose in living matter, another army of bacteria reduces the protein molecules of dead plants and animals to simpler compounds—nitrates and nitrites—that can be used again by plants. And finally, a third group of soil inhabitants, called the denitrifying bacteria, breaks down the nitrogen compounds by catalytic reactions, liberating gaseous nitrogen, which returns to the air.

Thus the passage of the nitrogen atom between the realm of living matter and the realm of the atmosphere is expedited by the enzymes, which guide each molecule across a secret passage between the two realms. Carbon atoms and the treacherous oxygen atoms are guided by similar enzymes. There is much that is still unknown about the way in which nature performs these miracles of transfer and how it responds to changes that might disturb the equilibrium. These gaps in our knowledge are particularly disturbing in view of the fact that human activities are making very significant changes in the quantities of chemicals present in the soil and the air.

But, in the absence of any interference from man, these wonderful natural processes connect the four portions of the biosphere, recycling all wastes and maintaining the precise combination of chemicals best suited to the support of the living matter that exists on Earth today.

— • —

Of course, none of these processes could take place at all without that most important of all the factors in our environment—the presence of water in the liquid state. Our planet is unique among those in the solar system; nearly three-quarters of Earth's surface is covered with water. Taking together the seas, the lakes, the rivers, the soil and ground water, the air with its film of clouds, we can think of the planet as being bathed in a warm, wet matrix for living things. Most of this water is in the seas; fresh water comprises only 3 percent of the total, three-fourths of which is locked up in ice. Almost all the rest is found in lakes and ground water. Only one-thousandth of 1 percent is present at any one time in the Earth's atmosphere. And yet this tiny proportion is responsible for almost all of the phenomena that we associate with weather—the swirling whiteness of winter snow, the gentle soaking rains of spring, the ice storms that turn every tree and bush into diamonds for a day, and the whirlwinds that descend with sudden violence out of a summer sky to toss mobile homes about and play havoc with fields of newly ripened grain.

But without the gift of water from the atmosphere life would not be possible on land. Water is the principal ingredient of living tissue. Our bodies are three-fifths water, and similar proportions are characteristic of most animals and vegetation. Water acts as a transportation system for many of the other chemicals that comprise living tissue. Most of these chemicals are soluble in water, and can be conveniently carried by this fluid medium from cell to cell, assuring a continuous stream of nourishment and energy.

Water has several special properties that are very advantageous in its role as a matrix for living things. It is a liquid within the range of temperature most suited to life processes. In this liquid form it changes temperature very slowly. More heat is needed to raise the temperature of a given quantity of water than is necessary for any other known liquid. Conversely, as heat is dissipated, the temperature of water drops very slowly. This gradual warming and cooling characteristic produces a remarkably stable and benign environment. The proximity of water in lakes and oceans and in the atmosphere helps to modify the temperature extremes that occur on land.

Like all other liquids, when water is cooled it contracts, but just before it actually solidifies into ice, water does something very unusual—it expands. Because of this strange property, ice is less dense than water; it remains on the surface of lakes and rivers. The shield of ice serves to protect the deep layers from contact with the colder air. Therefore, only very shallow bodies of water freeze solid and, in larger ones, aquatic organisms continue to survive beneath the ice.

Evaporation separates water molecules from the various other chemicals that water carries in the liquid state, so water vapor is very pure. Salt and all the other fellow travelers that are present in sea water are left behind when water is evaporated from the oceans, and the transformation of water into vapor is an efficient purifying process. Later, as water vapor condenses into rain, pollutants present in the atmosphere can dissolve in it, be carried down, and deposited on the land and sea.

— · —

For many years meteorologists were baffled in their attempts to answer a fundamental and seemingly simple question: How do raindrops form? Raindrops are at least a million times bigger than cloud droplets. Both theory and laboratory experiments showed that cloud droplets would not tend naturally to come together into larger drops.

The process of freezing proved to be the clue that solved the puzzle. As temperatures drop below 32 degrees Fahrenheit, most of the cloud droplets remain in the liquid state. In fact, temperatures can fall very far below "freezing" without any crystallization taking place and the moisture-laden air is said to be supercooled. Occasionally, however, one in a million of the droplets does freeze and once the tiny ice crystal exists in the supercooled cloud, water vapor molecules are attracted to it. Building outward from the center, molecules are added one by one. Each molecule attracts others and the pattern is built up symmetrically in ever widening circles. All of this occurs so rapidly that, to our senses, it appears to be instantaneous. Each crystal is an original masterpiece of lacelike perfection; matter has suddenly leapt from what seemed to be empty space and arrayed itself in form.

As the single ice crystal grows it becomes heavy enough to fall through the cloud, colliding with cloud droplets. Tiny splinters break off the crystal, leaving a trail of ice nuclei to start the formation of new snowflakes. These grow and fall and splinter, spawning more flakes; and so a snow storm blossoms from the presence of a single crystal.

If the snowflakes that are formed at high altitudes encounter warm air as they fall, they melt and turn into raindrops. Rain falls more efficiently than snow because its shape is more compact and it is denser. On their downward path the raindrops grow by collecting many of the little cloud droplets that they encounter.

Thus the formation of the first ice crystal in a cloud has started the process of precipitation. But this explanation just pushes the mystery back one stage. It is still necessary to explain how and why that first crystal forms. In 1942, two American scientists, Vincent Schaefer and Irving Langmuir, discovered the answer to this question. They were investigating the cause of ice formation on airplane wings, and in the course of this research they found that many clouds at high altitudes are supercooled. Anything passing through these clouds can act as a trigger to start freezing the water in the cloud droplets. Schaefer followed up this discovery with laboratory experiments. He found that by breathing into a freezer he could make miniature supercooled clouds. Then by introducing a very cold metal rod or particles of dry ice (frozen carbon dioxide), he could turn the cloud into snow and ice crystals. In November 1946, Schaefer conducted a field experiment to test his theory. Flying in a small plane, he identified a supercooled cloud at 14,000 feet over Schenectady, New York, and scattered six pounds of dry ice into its center. The cloud, approximately four miles long, turned into a snowstorm. One of Schaefer's colleagues suggested that the same triggering action might be achieved by using tiny crystals, which resembled ice in form but which did not reduce the temperature of the cloud. Silver iodide crystals were tried, and—just as postulated—the whole cloud turned into snow.

The important thing about a trigger reaction of this kind is that extremely small quantities of the added material can result in impressively large effects. Langmuir estimated that a single dry-ice pellet the size of a green pea falling through a cloud two-thirds of a mile in thickness could produce a hundred thousand tons of snow.

This trigger effect makes it possible for man to dream of controlling the weather. By scattering a few pounds of crystals, rain could be encouraged to fall on areas suffering from drought. It might even be possible to channel and divert the enormous forces that build up in the deceptively delicate-looking clouds that circulate around our planet. The basic principles seem simple at first glance, but when applied to an actual weather situation they are found to be extremely complex.

The fact that small modifications can precipitate large-scale changes in the atmosphere can be a disadvantage as well as an advantage to man. Over the past century, human activities have brought about inadvertent changes and added many triggering substances to the Earth's atmosphere. Thousands of tons of dust and ash pour forth every day from industrial smokestacks and exhaust pipes of automobiles. Many of these tiny particles can act as nuclei for condensation, just as the silver-iodide crystal does. The smog and haze that have become characteristic of cities are caused by condensation of water on these small particulates in the atmosphere. In air polluted by industrial waste, there are often as many as a million particles per cubic inch. Weather records provide strong evidence that the presence of this large number of particles is altering the weather.

Sunshine received on the Earth's surface decreased 1.3 percent or about eight minutes a day in the continental United States between 1964 and 1970, and cloud cover has increased. At La Porte, Indiana, the average annual precipitation is 50 percent greater since the installation of steel mills, foundries, and refineries at Gary, upwind of La Porte.

The La Porte situation was first documented in the early 1950s. About twenty years later, a much more ambitious study of the way cities modify the weather was undertaken in the St. Louis area. This city was chosen as the subject of the research project because it was believed to be large enough to cause significant effects and is located in the center of relatively flat land far from any large body of water. Measurements could be conveniently made in a wide circumference around the city. For about five years extensive data were collected on temperature, rainfall, hail, sunlight, and the levels of various pollutants. The results were very interesting. The rainfall in an area ten to thirty miles east of St. Louis (in the direction of the prevailing winds) was 20 percent higher than the rain in the city itself or in the outlying regions to the west, south, and north. Furthermore, the additional precipitation came mostly in the form of heavy downpours of one inch or more.

It is reasonable to conclude that these anomalies are caused by the presence of the city. The temperature in St. Louis was found to average 3 to 5 degrees above that of the surrounding countryside. Concrete and masonry absorb heat efficiently during the daytime and release it very slowly at night—much more slowly than a similar area of bare ground or ground covered by vegetation. This island of heat causes a rising column of air over the city and, since it is heavily laden with man-made particles, there are plenty of nuclei to trigger the formation of rain in the resulting clouds. These city-made clouds tend to rise to greater heights than those generated over the surrounding countryside. They seem to breed more violent storms and increase the incidence of hail.

From the results of this study in one typical American city, it is clear that inadvertent climate modification is probably very widespread. Much more information needs to be collected in order to understand in greater depth the effect of man's activities on cloud formation and on the distribution of water around the world. For example, the rate of evaporation of water from the land is being reduced every year by the paving of large areas of the Earth's surface. On the other hand, an increasing amount of water is being distributed by irrigation projects to dry desert areas where evaporation takes place very rapidly. Do these changes balance each other? We really do not know. The slightest alteration in the relative rates at which water is drawn up from the land and the sea and distributed again as rain and snow can cause very important changes in our environment. They can flood our streams and drown our river towns, or parch the lands that are usually lush with new growth in the spring.

—— • ——

City dwellers who have never lived through a major drought on a farm have only a faint hand-me-down impression of the long drawn out frustration, the feeling of helplessness, the intense longing for rain. Day after day the Sun shines in a cloudless sky with merciless and unremitting strength while the corn stalks wither in the fields. The soft, black soil turns hard and pale. The furrowed fields, which just a little while ago held rows of tender new plants, are now all gray, and the wind teases the bare soil into little dust devils that spread out and fill the air with a brown haze. Tantalizing clouds occasionally drift overhead and a thin veil of moisture falls but does not reach the ground. The hot, thirsty air has swallowed it whole. Sometimes a sudden shower passes by and the rain can be seen dampening a neighbor's fields but leaving yours dry.

Finally on a late summer evening you hear a faint murmur in the sky. Is it thunder? You have almost forgotten the sound. You wait, holding your breath. It comes again— louder, a roll that reverberates like music across the parched land. A dark cloud blots out the western sky, and the first raindrops begin to fall. Now a misty coolness softens the air like an embrace. And you lie awake half the night listening to the blessed sound of rain on the roof.

An instinctive recognition of the fundamental importance of water flows deep within all of us. Many people hold water sacred. They are baptized in it; they pour it over their heads at sunrise; they touch it to their foreheads in prayer. The sound of waves breaking on a beach or the pounding of rain on the roof is as reassuring as a mother's voice is to her child. Water is there, providing our nourishment, quenching our thirst, refreshing our bodies which would wither and die without it. Our bodies are three-fifths water that is salty like that of the ocean, perhaps a biological memory of the time when the first life forms were born and nurtured in the sea.

— • —

The sea was the cradle of life and yet, ironically, its movements can sometimes cause the winds and the clouds to withhold this most precious element—rain.

The oceans act as a giant moderator of the climate, warming air masses in winter and cooling them in summer. In fact, the upper ten feet of water contain as much heat as the entire overlying atmosphere. Ocean currents, and the rate of change of the thermal structure of the upper layers, are almost ten times slower than their atmospheric counterparts. Thus any abnormal temperature characteristics of the sea persist long enough to affect air masses moving over it. The gradient between unusually warm and cold pools of water is transferred to overlying air, producing zones of temperature differences that can alter windflow patterns. These gradients are rapidly transferred to higher layers of the atmosphere and ultimately to the jet stream, producing great meanders, which change the airflow paths over the continents. In this way they may produce persistent abnormal weather conditions such as the California drought of 1976 and the great Midwest drought of 1988.

Off the coast of Peru there occasionally develops a pool of unusually warm equatorial water extending westward to the central Pacific. Deviations may be as much as 5 degrees Fahrenheit above normal and may last for many seasons. This phenomenon has been called *El Niño* (after the little Christ child, because it often begins just before Christmas time). During these episodes, air over the tropics receives more heat and moisture from the ocean, so the air rises more rapidly than normal. Spreading poleward, it increases the momentum of the westerly winds and causes the major jet streams to move in deep loops across North America. As the El Niño episode subsides, the warm waters gradually move north of the equator and colder waters take their place (this current has been christened La Niña). Then the region dominated by rising tropical air occurs at 10 to 20 degees north latitude instead of near the equator. This change causes the jet streams and their associated band of high- and low-pressure systems to shift in a northerly direction as well.

The precipitation that these systems usually bring to the farm lands of the United States falls across Canada instead. This, in very brief form, is the theory of how El Niño is responsible for the droughts that have frequently caused the crops to fail in one of the great food-producing regions of the world. But what causes El Niño? Here is another of the many still unanswered questions about the Earth's weather system. The movement of ocean currents is at least as complicated as that of the air, depending on the temperature gradients, the position of land masses, and the rotation of the Earth. One possible cause of the temperature changes has just been discovered in very recent years—the presence of great rifts in the ocean floor where the Earth's crust is gradually pulling apart. In sections of the rifts, springs of extremely hot effluent are boiling up from the Earth's mantle, similar to volcanoes at the bottom of the sea.

The most active of these regions is on the East Pacific Rise south of the tip of Baja California and extending further south almost as far as Easter Island. Geysers of water as hot as 700 degrees Fahrenheit shoot up with great force twenty or more feet high. They are laden with sulfur compounds and many minerals in solution: copper, zinc, lead, and silver ores.

Submersible craft have been used to explore these hidden depths. The American craft named Alvin can descend to about 12,000 feet. It is equipped with long, mechanical arms to pick up samples from the ocean floor and stereo cameras that provide a continuous panorama of the wild, mysterious landscape of the rift valleys where jagged, black, lava mountains rise out of limestone sediment formations, soft and white as drifts of snow. This landscape has been described as otherworldly—perhaps more closely resembling Antarctica than any other scene on the crust of the Earth.

A weird assortment of aquatic creatures inhabits these

underwater mountains, clustering thickly in the vicinity of the hot springs—previously unknown species of fish, clams, mussels, lobsters, sponges, sea cucumbers, and worms. The water in the jets is thick with bacteria of a special kind that extract the energy stored in hydrogen sulfide. This smelly compound holds the secret of the abundant life in these regions, which are entirely cut off from the energy of sunlight. Under extreme heat and pressure in the Earth's mantle the sulfate normally present in seawater is converted to hydrogen sulfide, a compound in which heat energy is stored. The bacteria use the hydrogen sulfide combined with carbon dioxide and oxygen dissolved in the water to synthesize the organic compounds that are food for higher life forms. This unusual life-support system (the only known one not dependent on the energy of the Sun) is so successful that the living community is profuse and the species grow at extraordinarily rapid rates compared with other aquatic organisms. Giant red-blooded clams cluster in a dense blanket on the ocean floor; many of the rocks are almost solidly encrusted with enormous mussels as large as a man's head. In one area reddish fish of an unknown species spend most of their time with their heads thrust into the sulfur vents, feeding and moving their tails in unison like a flickering red flame. In another location, tube worms six feet tall grow tightly clustered together and open their spectacular red, flowerlike plumes to comb the nutrient-rich waters. The research team in Alvin called this hot spring "The Garden of Eden."

The hot springs are believed to be caused by sea water circulating through the outer layers of the Earth like a giant organism breathing out and breathing in. Cold water percolates down through porous rocks to the heated rock of the Earth's mantle. Then, scalding hot, it expands and pushes its way up again with great force. As it passes through the rocks of the crust it reacts with them, carrying along with it a load of dissolved minerals and gases. Methane gas and globules of oil have been found to be welling up from these vents.

The quantity of extremely hot water emerging from the rifts is the most astonishing fact. Scientists have estimated that sea water is pouring through the system at a rate of something like a million and a half gallons per second, and the water in some of the vents is hot enough to melt the instruments sent down to measure the temperature.

It would not be unreasonable to suppose that the influx of these waters could create warm currents on the surface of the sea. And as we know from volcanic activity on land, volcanism is not a steady phenomenon. It comes in waves and then passes through quiescent periods. In a similar way, El Niño comes and persists for several seasons and then may not appear again for a number of years. The vents on the East Pacific Rise apparently turn off after a decade or two and are replaced by other vents elsewhere along the seafloor rifts.

Any connection between El Niño and the ocean vents is very speculative, however. The underwater activity has been discovered so recently that its true extent and significance have not yet been determined. It would be interesting but not really surprising to find that volcanism 10,000 feet below the surface of the Pacific can create a summer without rain in the Midwest, just as the eruption of Tambora in Indonesia caused a year without a summer.

Mankind has only recently begun to appreciate the intricate way in which all the natural systems are interwoven, creating an elaborate fabric out of a few basic threads. It is hard to move even one tiny part without disturbing all the rest. As John Muir expressed it: "When we try to pick out anything by itself we find it hitched to everything else in the universe."

Unlike human affairs, where size or quantity or power are

the measure of value, in nature the importance of an element cannot be judged by magnitude alone. A handful of silver-iodide crystals, microscopic grains of dust, few tiny bacteria living deep with the soil or under the sea—such small things can cause the heavens to open, can move the swift jet streams, and make life flourish or wither across the great continents of the Earth.

In so many ways nature is telling us to look for answers in subtle processes and balancing acts so fine they can be disturbed or restored by a magic molecule or two. Witness the artistry that nature has lavished on the design of even the littlest things—the iridescent wings of the dragon fly, the delicate pattern of the limpet shell, the shining perfection of the snow crystal. What better way of telling us that small is beautiful?

4

·

The Kingdom of the Sun

Solar corona photographed during a total eclipse. (Courtesy of the High Altitude Observatory, National Center for Atmospheric Research.)

The presence of life on Earth changed the planet. It softened the hard crust of the continents and altered the composition of the atmosphere; it helped to create the series of halos that surround it like the protective membrane that encloses each living cell, filtering out harmful elements and passing through those that are needed for energy and growth. "When the earth came alive," Lewis Thomas said, "it began constructing a membrane for the purpose of editing the sunlight."

It is the Sun that dominates the upper layers of the Earth's atmosphere. But ironically it is photosynthesis (the use of the Sun's energy to make food for life) that has created the layer that prevents the most energetic rays of the Sun from reaching the Earth's surface.

The presence of this ozone layer was discovered when balloon and rocket flights began bringing back information about the atmosphere high above the realm of the clouds and the weather. The temperature change at the tropopause, occurring about six or seven miles high in midlatitudes and extending to altitudes of about thirty-two miles, was a strange phenomenon that needed explanation. In this layer the atmosphere is extremely rarefied; its density is less than one-thousandth of the density of air at sea level. Studies of these tenuous gases revealed the presence of significant amounts of ozone (the three-atom molecule of oxygen). These atoms are created by sunlight acting on free oxygen in the upper atmosphere. During the day, the Sun breaks down some of these molecules to single oxygen atoms and these react with the oxygen molecules that have not been dissociated, to form ozone. However, the sunlight also breaks down ozone, converting some of it back to normal oxygen. And naturally occurring nitrogen oxides enter into the cycle, speeding the breakdown reactions. The amount of ozone that is present at any one time is the balance between the processes that create it and those that destroy it. The net amount is small. About eight in a million molecules present in the stratosphere are ozone. If it were all brought down to sea level, where it would be subjected to the atmospheric pressures we usually experience, it would occupy a layer only one-eighth of an inch thick. So thin a shield protects the whole biosphere—hardly more than the thickness of an umbrella!

Although such a minor constituent of the stratosphere, ozone is responsible for the rising temperature at this level. It absorbs the ultraviolet radiation from the Sun, converts it into heat and chemical energy. The difference in temperature caused by the presence of ozone inhibits vertical air circulation at the level where this inversion occurs and thereby creates the tropopause, which divides the lower atmosphere into the troposphere and the stratosphere, which contains the ozone layer.

Above the tropopause, a particularly beautiful cloud formation can sometimes be seen just before sunset and as long as three hours afterward. These mother-of-pearl clouds are composed of ice crystals, iridescent and brilliantly banded with color. They are most often observed in extreme northern regions—Alaska and Scandinavia—when the air is exceptionally clear. They are believed to be triggered by dust or other fine particles that have reached the lower stratosphere, perhaps from a major volcanic eruption.

This hypothesis has been substantiated by measurements made of volcanic dust clouds. After the eruption of El Fuego in Guatemala in 1974, the dust cloud was tracked by NASA scientists using a very accurate system known as lidar. Pulses of laser light beamed at the cloud were reflected back to Earth, and the time for this round trip was very precisely measured. It was found that shortly after the eruption there were two layers of clouds at approximately nine and twelve miles. Months later the clouds had joined into a single cloud with maximum density at eleven-and-one-half miles but extending up to fifteen miles, within the range of altitudes at which mother-of-pearl clouds have been sighted.

Throughout the stratosphere, air circulation is very sluggish. Elements introduced there may remain for years until they are ultimately transported downward into the lower levels and washed out in rain. The stratosphere has been likened to a city whose garbage is collected every few years instead of daily. Since there are no practical ways of speeding up the cleaning operation, we have no alternative but to live with several years' accumulation.

Horizontal mixing takes place more rapidly. Substances released in the stratosphere spread in an east–west direction around the world in a week, and from the equator to the poles within a few months.

Following the contour of the tropopause, the ozone layer lies closest to the Earth over the poles and rises to maximum height over the equator. Since the splitting of the oxygen molecules depends directly upon the intensity of solar radiation, the greatest net rate of ozone production takes place over the tropics. But horizontal circulation patterns carry the ozone-enriched air away from the equator. It is a surprising fact that the largest total ozone amounts are found at high latitudes. On a typical day the amount of ozone over Minnesota, for example, is 30 percent greater than over Texas, 900 miles farther south. The density and altitude also change with the seasons and the weather. However, until very recently, scientists believed that at any one place above the Earth's surface the long-term averages were reasonably constant. So it was shocking to discover that the ozone layer is becoming thinner around most of the Earth and that a hole in the layer develops every springtime over the South Pole.

The consequences of a depleted ozone layer are serious.

Life on Earth would be exposed to very damaging radiation. Furthermore, the temperature differences at the tropopause would be weakened, altering circulation patterns in the atmosphere. For these reasons, the Earth's weather system would undoubtedly be affected. How severe and how far-reaching would the changes be? We will consider these controversial questions after we have completed our tour of the outer layers of vapor that envelop the Earth.

— • —

Beyond the stratosphere are regions that are even more mysterious, responding in ways still only dimly understood to the power of the Sun and the other forces that flow in upon it from the farthest reaches of the galaxy.

Above the stratosphere with its layer of ozone lies the *mesosphere,* a large region (about twenty miles deep), which is almost invisible. However, a very rare type of cloud formation is sometimes sighted there. About a quarter of an hour after sunset clouds begin to materialize in a sky that had earlier seemed to be perfectly clear—between the twilight arch at the horizon where the Sun has gone down and the darkening blue vault of the sky overhead. By an hour later they show up very clearly, reflecting the sunlight with a silvery-blue glow that shades into golden yellow near the horizon. The delicate feathery ripples of foam look like waves in a phantom sea.

The altitude and speed of movement of these noctilucent clouds have been precisely measured. They always occur at heights of approximately fifty miles, above 99.9 percent of the atmosphere. This is the location of the mesopause, the thin layer of extremely cold atmosphere that marks the upper limit of the mesosphere. The temperatures (down to 225 degrees Fahrenheit) are so low that any water vapor present there would be frozen, but measurements of the size and reflectivity of the clouds indicate that dust as well as ice is present. What is the source of this dust? Does it come from Earth like the volcanic ash that triggers the mother-of-pearl clouds, or is it brought in by shooting stars and comets from outer space? In 1908, an enormous meteorite estimated at 1,000 tons fell in Siberia. This event was immediately followed by the appearance of very striking noctilucent clouds.

To answer these questions, an experiment was conducted in Sweden in 1962 by a team of American scientists working in cooperation with experts from the University of Stockholm. Their equipment consisted of four rockets, each carrying a device for collecting a sample of the cloud particles, a parachute for returning the sample to Earth, and a radio beacon to direct the researchers to its landing site. During the late summer of 1962, experienced observers throughout Scandinavia watched for these characteristic cloud formations. The whole population of Sweden was alerted by press publicity to report clouds that were visible long after sunset. The first sighting was made in early August and a rocket was fired into the cloud. The other rockets followed later that month. The particles collected and returned to Earth proved to be grains of dust that had been coated with ice. The size and chemistry of the particles indicated an extraterrestrial origin. They were considerably larger than the maximum size of dust from the Earth or volcanic ash that could have been carried to these heights. The participating scientists felt confident that these particles came from outer space, perhaps a wisp of tenuous matter from a comet's tail or fragments of a shooting star.

— • —

The farther we move away from the surface of the Earth, the greater the effect of the almost undiluted energy of the Sun.

But the influence of the planet is still making itself felt even at distances 400 miles high. A small portion of the gases produced here on Earth find their way up through the breaks in the tropopause, higher still through the slow circulation patterns of the stratosphere and the cold, dark spaces of the mesosphere, up to the place where they feel the maximum impact of solar radiation. The reactions that take place between the Sun's energy and the tenuous gases produce a diffuse glow, as though space itself were giving off light.

For many centuries, people had been puzzled by the presence of this faint but continuous yellow-green light like the cold phosphorescence of fireflies. On the clearest of nights when there is no Moon, the sky between the stars is very dark but not absolutely black like India ink. All the starlight added together is insufficient to produce the glow, which can be seen equally well from vantage points around the whole planet. Its symmetrical distribution suggests something like a spherical halo. But it is difficult to see a halo for what it is when one is inside looking out. The shine of it slightly hazes the outline of things beyond the glow, and it is hard to tell whether the haziness comes from the things themselves or from something in between.

By the early twentieth century, optical instruments had become available for analyzing the wavelength of radiation and identifying the elements that produce it. Oxygen and nitrogen as well as a number of less-common elements proved to be the source of the airglow. In a wide, diffuse band of atmosphere, intense solar radiation falls upon these gases and causes reactions that turn neutral atoms and molecules into ions (particles that carry a net electric charge). There are reactions that return the ions to their neutral state and many of these give off light. The emission continues long after the Sun has passed into the Earth's shadow. In the daytime the light emitted by these air molecules is stronger, but our eyes are so dazzled by other light that we cannot see it. From satellites orbiting in darkness outside the Earth's atmosphere, the airglow can be seen and photographed. This luminous halo extends from fifty miles to 300 or 400 miles high and is known as the *ionosphere.*

— • —

By far the most dramatic effects that take place in the ionosphere are the brilliant displays known as *aurora borealis* or *aurora australis,* the "northern or southern dawn." The aurora displays do not occur at dawn, however. They usually begin a few hours after sunset and are all over by sunrise. An observer in the Northern Hemisphere looking toward the dark northern horizon sees it begin to shine with a pale, greenish light. It brightens rapidly; white rays faintly tinged with magenta and lime green arch high in the sky. More and more rays of light appear; they move, break up, and re-form in wavering luminous sheets like firelight flickering on the ceiling of the world.

Throughout recorded history tales have come down to us describing the awe and wonder inspired by these nocturnal displays. In far northern lands where aurora borealis is almost a nightly occurrence, it became part of the folklore and religion of the people. Some tribes believed that gods carrying flaming torches were dueling in the dark heavens. To the Eskimos in Hudson Bay, the polar lights are cast by lanterns carried by demons who are searching the universe for lost souls.

But farther south where such displays are very rare, they caused great fear. The story is told that, in 1585, on a clear autumn night, thousands of terrified French peasants left their homes in the peaceful countryside and crowded into the churches to pray. The sky was on fire! Curtains of flame rolled in over the northern horizon and beams of white and red light shot high as from a gigantic explosion just over the edge of the world. Shortly after midnight it seemed as though the prayers of the people had been answered. The firelight in the sky slowly subsided and soon the dome of heaven was once more tranquil and studded with stars. Gratefully, the people left the sanctuary of the churches and crept back to their cottages.

— • —

Ground-based studies of the aurora, later confirmed and greatly expanded by information from satellites orbiting the Earth, has revealed the nature of these remarkable color and light shows. The Earth is immersed in a stream of charged particles flowing outward from the Sun. This stream is called the *solar wind* and, like the winds on Earth, it is variable, blowing sometimes very hard, sometimes more softly.

The solar wind consists mainly of ionized hydrogen— protons and electrons having positive and negative charges, respectively. It flows at tremendous speeds, between 220 and 500 miles per second, taking on the average about three and a half days to cover the distance from the Sun to the Earth. This solar wind is incredibly tenuous but very hot because of the high velocity at which the particles are moving.

Just as a current of air or water sweeps around any solid object in its path, the solar wind envelops the Earth. It interacts with the geomagnetic field in complex ways, set- ting up regions where charged particles are accelerated and then trapped in space. They oscillate back and forth until some eventually work their way out of the magnetic trap, and are funneled along the Earth's magnetic field lines into the polar regions where they strike the molecules of the ionosphere with great force, blasting them apart, creating excited molecules and more ions. When these ions return to a more stable state, they emit the light that we see as the aurora. The continuous bombardment of the ionosphere causes it to glow in two great luminous ovals centered on the magnetic poles—a phenomenon that has been seen and photographed by satellites orbiting the Earth. These are the sources of the shimmering sheets of light that are seen regularly at very high latitudes. When the solar wind is exceptionally strong the auroral ovals expand into much lower latitudes. The floating veils move across the night sky at supersonic speed, about 2,000 miles per hour in the low-density air of the upper atmosphere.

Auroral displays occur with the greatest frequency in a belt about 23 degrees from the magnetic pole in each hemisphere. Between 60 and 45 degrees—in Canada, most of northern Europe, and the British Isles—they are seen frequently, but below 45 degrees they are rare. The most brilliant displays are likely to occur at the times of the equinoxes—March and September.

Oxygen atoms produce most of the color effects. The commonest color is greenish white, a wavelength emitted by excited oxygen dropping back to a stable state. A red light is also produced by oxygen in a more highly excited state. A beautiful pink emission comes from excited molecules of nitrogen.

— · —

The fact that auroral displays have caused concern and comment over the years has resulted in an historical record of these events which is complete enough to identify the time and place of their occurrence over a span of several centuries. One of the most significant discoveries is that their incidence waxes and wanes in approximately eleven-year cycles. This fact provided one of the clues that suggested their origin. It had also been observed that certain mysterious dark spots on the Sun increased and decreased on a cycle that varied between eight and fourteen years but averaged just slightly over eleven. When these two phenomena were plotted together, a remarkably close correlation was found, and these facts led scientists to theorize that auroral lights are caused by jets of charged particles emitted by the Sun.

Chinese observers using the unaided eye had reported seeing sunspots more than a thousand years before they were definitely identified by scientists in the Western world. In 1611, after the invention of the telescope, Galileo studied these strange markings on the Sun and described them in a book that bore the title *History and Demonstrations Concerning Sunspots and Their Phenomena*. Unfortunately, Galileo's studies of the Sun damaged his eyes, which had already been impaired by an infection in his youth. During the last four years of his life he was completely blind.

The eleven-year cycle of sunspot activity was not noticed until the late eighteenth century and was not confirmed until 1843. As a matter of fact, the cycle itself did not occur in a regular manner during the hundred years after sunspots were described by Galileo. Throughout that century the Sun seems to have behaved very strangely. The old records of sunspots, which are very sparse and, therefore, not completely conclusive, show that two maximums may have occurred in the years immediately following 1611, the peak times coming about fifteen years apart. Then the intensity of sunspot activity declined and remained very low until 1715. During that time polar lights were extremely rare, even in such northern cities as Copenhagen and Stockholm, and when they did occur, they caused great consternation.

Throughout these seventy years, the paucity of auroral displays was only one of several indications of decreased solar activity. Normally during total eclipses of the Sun, a rim of flaming gases can be seen with streamers extending thousands of miles above the solar disk like the petals of a flower (see the photograph on page 70). This phenomenon, known as the solar *corona*, was not reported in total eclipses during the seventy-year period of the quiet Sun.

Another interesting piece of evidence has turned up in the study of tree rings. The size of tree rings tends to increase and then decrease again over a period of ten years or so. But samples of wood from the seventeenth century show little or no evidence of such cyclical changes in growth.

Although the relationship between sunspot activity and weather has not been firmly established, there are interesting correlations which suggest that such a relationship exists. The years of the quiet Sun were centered on the period known as the Little Ice Age. In 1550, the weather in Europe began to grow steadily colder. Throughout the next three hundred years temperatures in northern Europe were considerably colder than they are today. The Arctic ice pack was greatly extended. Paintings and engravings of that time show scenes of ice skating on the river Thames, which has not frozen over in recent times. The winter of 1695 was especially severe; famine took a major toll in Finland, Estonia, Norway, and Scotland. The year 1771 was the worst in a series of wet and snowy summers in central Europe; this was again a time of famine and the rapid advance of the Swiss glaciers. An engraving made about 1855 depicts the Argentière glacier near Chamonix in the French Alps as considerably more extensive than it is today. Evidence collected by the French historian Emmanuel Leroy Ladurie documents the fact that Alpine glaciers grew almost continuously between 1590 and 1850. Geologists have found evidence of glacial growth from Patagonia to Kanchenjunga, from Mt. Kenya to the Southern Alps of New Zealand. In the United States reliable records indicate that average midwestern temperatures in the 1830s were as much as 7 degrees colder than today.

These centuries of cool weather had been preceded by a long period of warmer climate. Historical records show that most of northern Europe enjoyed an unusually benign climate between the years 400 and 1200. The twelfth century, falling toward the end of this trend, coincided with a period of intense solar activity as evidenced by increased aurora displays and spectacular corona.

There is striking historical evidence of the changes in climate between the tenth and the fifteenth centuries. This information comes from the Old Norse sagas and from the remains of the Viking settlements in Greenland. Iceland was settled in the mid-ninth century by Norsemen. Among the colonists living there toward the end of the tenth century was Eric the Red, named for the color of his hair and beard. In 982 he was banished from Iceland for three years as punishment for murdering two men in a feud. Eric had heard a persistent rumor that more land lay beyond the sunset. Inspired by this tale, he sailed west and discovered a large island about 180 miles away. There is no mention in the sagas that he encountered ice drifts, which are now common in these northern waters. During his three years of banishment he explored this land and then returned to Iceland with stories of lush landscapes and fertile pastures in the island that he called Greenland.

Eric, a skillful promoter, believed that more people would follow him to the new land if it had a beautiful name. Greenland was probably never very green. But at that time it did have a climate benign enough to grow vegetables and grasses to make hay for livestock.

Attracted by these reports, about 300 men, women, and children together with cows, horses, sheep, and household goods set sail in open Viking longboats for new homes in Greenland. Here they established several colonies. They built houses, cultivated farms, bred cattle, wove cloth, hunted, fished, and traded with Europeans. There seems to have been no difficulty during this warm period of maintaining regular communication with the continent. In fact, more and more Viking families made the long journey from Scandinavia with their possessions to join the colonies, which prospered for at least a century or two.

In the thirteenth and fourteenth centuries, however, the communities decreased steadily in size. Travel to and from Europe became more difficult. Ships headed for Greenland had to go farther south to avoid the drift ice and then swing back north to reach the southwestern coast settlements. Finally, all communication with the homeland ceased.

Excavations of the ancient farms and grave sites show many farms had been abandoned; fewer cattle, sheep, and horses were raised. The human skeletons found in the later graves were small in stature; by about 1400, the average Greenlander was probably less than five feet tall. Many seemed to have been severely crippled and twisted. The cause of this condition has not been definitely determined— disease, poor nutrition, or perhaps decalcification of the bones after death.

Most striking of all is the fact that the graves became shallower as time went on. They were dug six-feet deep in the twelfth century but gradually decreased in depth until by the end of the fifteenth century they were barely deep enough to cover the bodies. These facts can be most easily explained as a result of increasingly colder weather, making it more and more difficult to dig into the permanently frozen ground. Finally, in the early sixteenth century, no colonists were left in Greenland. Today, the site of Eric the Red's settlement is once again green; sheep graze there and the glaciers have melted back to open the fjords.

— • —

These and a number of other correlations have been discovered between the climate on Earth and the dark markings on the Sun. We do know that the sunspots are an indication of

increased solar activity. Flares and eruptions are concentrated in regions near the spots where magnetic field lines
meet, and these magnetic fields reverse their polarity every
eleven-year cycle; thus, every twenty-two years the same
situation is repeated. But measurements (in space) of solar
energy striking the Earth's atmosphere during the peak periods show a change of approximately .16 percent over the
average cycle. Although some scientists are skeptical that
this small a change could have an important effect on the
weather, there is a growing consensus that a statistical correlation exists between the occurrence of certain weather
patterns and the eleven-year sunspot cycle. For example,
droughts on the Great Plains have occurred on approximately a twenty-two-year cycle, coinciding with every other
solar minimum.

Eugene N. Parker, Professor of Astronomy and Physics at
the University of Chicago, has made a study of the Sun his
specialty for many years. He believes that there may be a
consistent relationship between temperature on Earth and
solar activity. "When the Sun is in an active phase," he says,
"it produces high levels of ultraviolet radiation and x-rays.
The increase of solar energy involved is only a very small
fraction of the total, but it is sufficient to heat the tenuous
gases of the upper stratosphere and ionosphere as much as
20 to 30 percent. And this warming seems to be associated
sometimes with winds that carry warm tropical air toward
the poles. It is not unreasonable to suppose that changes of
this magnitude might in some way affect temperatures here
on the surface of the Earth. So far, however, careful monitoring for such variations has not produced consistent evidence
that this is so." We are presently (in 1989–1990) experiencing
an unusually high solar maximum, the largest that has been
recorded since 1958. Using measurements from satellites,
the search over the next few years may resolve this question.

— • —

Temperature changes much greater than the Little Ice Age have descended on our planet, and no one believes these could have been caused by variations in the energy output of the Sun. They may have been triggered by differences in the way this energy was received on Earth.

Throughout geologic history there have been sweltering "summers," lasting for millions of years. On the other hand, there have been bitter "winters" that held the planet in their icy grip for tens and hundreds of thousands of years.

The record of the temperature variations has been preserved in ice and in the crust of the Earth. Glaciers have left their characteristic grooves in very old rocks, and sediments bear telltale marks of changing sea levels. Fossils provide clues to the conditions that prevailed when these organisms were alive. These are some of the many bits of information that geologists use to understand the planet's ancient history.

According to this record the Earth has passed through five major glacial epochs. These large cycles of climatic change have occurred at very widely spaced intervals—at least 200 or 250 million years apart, with long periods in between when the Earth was warm and free of ice. Superimposed on these glacial epochs are many smaller fluctuations—briefer ice ages that waxed and waned in periods of about 100,000 years. These are the ones that have the greatest relevance to the human condition. We are living in one of the few warm periods that have occurred during the last million years (see Figure 2).

It has been possible to reconstruct in considerable detail the most recent sequence of shorter ice ages by studying the chemistry of, and organic remains in long cores extracted from the ocean bottom and glaciers as well as from the land itself.

The element oxygen provides important information. Most oxygen atoms have an atomic weight of 16 but some are a little heavier, having a weight of 18. Because oxygen-18

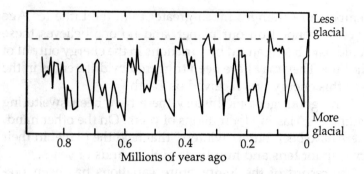

Figure 2. Global ice volumes of the past million years based on oxygen isotope measurements in deep-sea cores. The dashed line shows that times with as little ice as at present have been rare and brief. (Adapted from Reid A. Bryson and Thomas J. Murray, Figure 10.1, in *Climates of Hunger,* by permission of University of Wisconsin Press.)

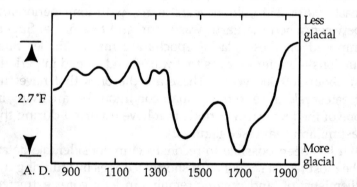

Figure 3. Northern Hemisphere temperatures of the past 1,000 years. (Adapted from Reid A. Bryson and Thomas J. Murray, Figure 10.1, in *Climates of Hunger,* by permission of University of Wisconsin Press.)

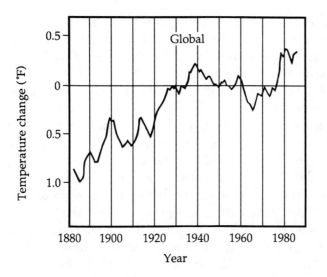

Figure 4. Global surface air temperature trends of the past 100 years constructed from averages of the 5 years centered on the plotted year. (Adapted from graph published by NASA Goddard Institute for Space Studies, 1988.)

is heavier, it does not evaporate as easily and so makes up a smaller proportion of the oxygen in clouds, rain, and snow than is found in sea water. In an ice age the snow, with its smaller proportion of oxygen-18, is sequestered in the ice caps. The oceans, on the other hand, are richer in oxygen-18 than at other times, and this shows up in the shells of tiny marine organisms. The presence of fossils whose age is known helps in dating the various levels of the core. These methods are effective over the past half-million years.

During this time the Earth's climate has oscillated between ice ages and warmer interglacial periods. The cold periods have lasted much longer than the warm ones. Ice ages averaging about 100,000 years in length have actually dominated the Earth's climate as long as human beings have occupied the planet. The warm interglacial periods have been short—typically 10,000 years in duration. The one we are enjoying now has lasted for at least 9,000 years.

Many theories have been proposed in an attempt to explain these severe fluctuations in the planet's climate. The cause of the major ice epochs is still a mystery, but the smaller variations throughout the last million years may be explained on the basis of regular changes in the amount of solar radiation received at the planet's surface.

In the 1920s a Yugoslavian scientist, Milutin Milankovitch, drew the attention of the scientific world to periodic variations in the Earth's movement around the Sun. Following up on this work, a German climatologist suggested that resulting changes in the distribution of incident solar energy could account for the ice ages. At first this theory aroused only passing interest because, at that time, it was believed that there had been only four recent ice ages, and the timing of these did not correspond with Milankovitch's theory. But more accurate information from the study of ocean cores has revealed the fact that there have been many ice ages— perhaps twenty—within the last 2 million years and their spacing does seem to correlate remarkably well with the cyclical changes in the shape of the Earth's orbit, the tilt and precession of its axis as it moves around the Sun.

The Earth's orbit varies slightly from a true circle; the ellipticity changes slowly from year to year, becoming more elliptical and then less so in a cycle of 92,000 years. When the orbit is almost perfectly circular, the amount of solar energy intercepted by the Earth is nearly constant throughout the year. As the ellipticity increases there is a small seasonal variation in the radiation that reaches the planet.

Seasons on Earth, however, do not depend on this very slight divergence in distance from the Sun. A much larger factor is the tilt of the Earth's axis, causing different portions of the planet's surface to face toward or away from the Sun. The warmest weather occurs near the equator, where the

Sun's rays follow the shortest path through the atmosphere. At higher latitudes the radiation passes through greater amounts of atmosphere and strikes the surface at a glancing angle. So the effectiveness of the Sun's rays is reduced.

The tilt of the Earth's axis changes slightly from year to year, swinging in a long cyclical pattern from 22.1 to 24.5 degrees and back again in 40,000 years. This variation in the angle of tilt has an effect on the degree of seasonal change.

A third factor was suggested by Milankovitch. There is a wobble in the Earth's axis due to gravitational interactions of the Moon and the Sun on the Earth's equatorial bulge. This wobble causes the planet to precess like a spinning top. The angle amounts to about 1.5 degrees a century, and it completes a full circle in 21,000 years.

These three periodic variations affect in small but significant ways the amount of solar energy that reaches each point on the Earth's surface. Contrary to what one might expect, the times of maximum eccentricity and maximum tilt provide the least favorable conditions for glacial buildup. At these times, the greater seasonal differences result in summer temperatures warm enough to melt the snows accumulated from the previous winter.

The three Milankovitch cycles set up a complex but predictable rhythm, sometimes reinforcing, sometimes canceling each other. When their combined effect was plotted, an impressive correlation was found between this curve and the geologic record from the study of oxygen isotopes and fossil marine organisms in ocean cores. For this reason, the Milankovitch hypothesis is quite generally accepted today, although there are some arguments about the fine points of the timing and a few geologists who are not totally convinced.

All in all, however, this seems to be our best model today for the prediction of changes in the next several thousand years. We have just passed a peak in the Milankovitch curve and are headed downhill toward another ice age. How rapidly will the effects be felt? It has always been assumed that the descent into an ice age occurred in a slow and ponderous fashion. But disturbing word comes from climatologist Reid Bryson, who has made a detailed study of the record. He finds that climatic change comes more abruptly than we have imagined. Shifts from glacial to interglacial have occurred within a century or so (see Figure 2). Shifts in the other direction take place a little more gradually because glaciers build up more slowly than they melt. However, the record also shows a relatively rapid descent into ice ages.

— · —

It is not surprising to find that the Sun rules our lives, that it holds the power to strip away the glaciers that guard our coastal cities from flood or precipitate us into the prison of a new ice age. The Sun is a despotic ruler; its energy is so great that we cannot afford to look directly upon its face. A veil must be thrown up to soften its piercing gaze. And if through carelessness we tear this veil, we may become blind or dead.

And yet what joy there is in the presence of the Sun! Its warm touch feels like a caress. When it hides from us for many days, the hours are grey; we are depressed and sad. Then suddenly the clouds part and the Sun shines down upon the Earth, gilding the green fields, intensifying the blue of the sky, and making a million lights dance on the surface of the sea.

At such a time it is easy to understand the almost universal belief in a Sun-god. We can imagine greeting the rising Sun with the Druids on a summer solstice, or worship the Egyptian Aton symbolizing the Sun's gift of life to the world, or watch with the ancient Greeks while Apollo drives his golden chariot across the sky.

PART II

·

ASSESSING THE DAMAGE

5

— • —

Hothouse Earth?

Fire followed by erosion can destroy the land. (Photograph by George E. Griffith, courtesy of the U.S. Forest Service.)

Hawaii is one of the youngest islands in the world. It began boiling up from the sea about half a million years ago and is still being created. Of the five great volcanoes on this island, two are still active. Mauna Loa and Kilauea occasionally spurt forth fountains of fire. Lava pours out of glowing vents on their flanks and flows in long ribbons of fire down toward the sea. The island is being made before our very eyes; as each eruption adds more substance, the volcanic cones grow ever larger.

Even while the island is being formed, life has clothed it in a mantle of green. The steep mountainsides that are not covered with fresh lava are softened with a cover of vegetation, so uniform and dense it looks like moss. The lower slopes are more variegated, layered in different shades of green—the soft understory of fern trees, the delicate leaves of the *'olapa* trees and the tasseled tops of lobeliads are shadowed by the dark canopy of ironwood and banyan trees. Drifts of pale *kukui* trees flow like the rivers of lava down long lush valleys to the sea.

It is very fitting that here in this place where the creativity of nature is so stunningly displayed, where the fertility of volcanic soil and the most gentle of all climates combine to make a little paradise, an observatory has been built to monitor the atmosphere's vital signs. It aims to measure continuously the changes that mankind is making in the texture and chemistry of the air, as the effluents from 100 million smoke stacks, the dust from drought-stricken fields, the exhaust from a billion cars are released into the atmosphere and sown on the wind.

Mauna Loa is as favorable a location for monitoring long-term changes as any that can be found at midlatitudes. Except during volcanic eruptions when plumes of vapor are emitted, this location provides a relatively pollution-free environment, far from the regions most affected by the activities of man. The air, borne in on the steady northeast trade winds, has traveled over thousands of miles of open sea.

During the last thirty years, records have been kept at Mauna Loa on the concentration of carbon dioxide in the air, and it has been steadily rising. Carbon dioxide is one of the trace gases in our atmosphere; it represents approximately one-third of 1 percent or about 300 to 400 parts per million. In 1958, the data from Mauna Loa indicated an average level of 315 parts per million, and this increased a little every year until, in 1989, the concentration was approximately 350 parts per million. This steady buildup of carbon dioxide is believed to come primarily from the burning of fossil fuels—a phenomenon that is increasing every year as industrialization spreads to underdeveloped nations and the human population continues to grow. These facts have caused worldwide concern about their effect on the Earth's climate.

Elements present in the atmosphere—even the very dilute trace elements—can change the way in which solar energy is transmitted to the Earth and the way the planet's energy is radiated into space. Earth's surface, whether land or water, emits infrared (long-wave) radiation; half of it is in the wavelength interval between nine and seventeen micrometers, where the atmosphere is normally quite transparent. This interval is called the "atmospheric window." But carbon dioxide and several other trace gases absorb radiation of these wavelengths very strongly and re-radiate it back toward the surface of the planet, thus reducing the transparency of the window by which infrared radiation escapes into space. Short-wave radiation, on the other hand, like the energy that flows from Sun to Earth, can pass uninhibited through this window. Like the glass in a greenhouse, the window allows energy to flow in one direction—from the stratosphere to the Earth—but does not allow it to pass as freely in the other direction—from the Earth to the stratosphere. Theoretically this results in a net warming of the planet's surface and a corresponding cooling of the stratosphere.

The theory of the "greenhouse" effect has been well established for almost a century, but until recently there has not been any clear indication that the burning of fossil fuels has actually had a detrimental effect on the Earth's climate. Even now, the indications are not unequivocal.

Reliable records of temperature are available back as far as 1850, when the Industrial Age began. These records show that during the last half of the nineteenth century and up until the early 1940s temperatures on the average rose (see Figure 4). In spite of small fluctuations, there did appear to be a progressive change, although the warming during this time cannot all be attributed to higher concentration of carbon dioxide because the increase in levels of this gas began very slowly and did not reach appreciable amounts until well into this century. Reduced volcanic activity from the early 1920s to about 1950 may also have been a contributing factor. However, all in all the change appeared at the time to be beneficial.

During the long warming period the glaciers retreated; the length of the growing season in England increased by two to three weeks. By 1940, the coasts of Europe and Asia were open to shipping for an extended summer season. Ports in West Spitsbergen were free of ice for seven months of the year, compared with the three-month period at the turn of the century. Ice floes in the Russian portion of the Arctic Ocean decreased by 350 square miles between 1924 and 1944. During these years various species of birds and fish typical of warmer climates extended their range into higher latitudes. The black-browed albatross, the ovenbird, and the Baltimore oriole summered in Greenland; wood warblers, skylarks, and scarlet grosbeaks visited Iceland, while the real Arctic species moved farther north.

In 1912, the first codfish appeared off the coast of Greenland—a fish never before seen by the local inhabitants. By the 1920s cod fishing had become a major occupation and the center of the fishery was moving steadily north. Throughout this period the Northern Hemisphere enjoyed a benign and relatively stable climate.

But, then, in the late 1940s a reversal occurred, and average temperatures—at least in the Northern Hemisphere—turned colder. (Records for the Southern Hemisphere were not sufficient to document the trend there.) The most pronounced temperature changes occurred in the high latitudes. This variation by latitude is typical of changing climatic patterns. The greatest alterations occur toward the poles, while in equatorial regions there may be little or no change. The average temperature in Iceland, for example, was 4 or 5 degrees colder in 1973 than it had been around 1940. The Northern Hemisphere as a whole cooled somewhere between .5 degree and 1 degree Fahrenheit. According to Hubert Lamb, director of the Climatic Research Unit at the University of East Anglia, the growing season in England dropped back by about two weeks during that time. The temperature of the western North Atlantic declined 3.6 degrees Fahrenheit between 1950 and 1970.

Climatologists began to warn that we might be going into a new ice age. Their concern was heightened by the knowledge that the orbital variations of the Earth as described by Milankovitch predicted declining temperatures in the Northern Hemisphere for the next 20,000 years. And, although this seemed to be too long a time span to hold any relevance for human affairs, there was evidence that the descent into an ice age could occur in just a century or so.

Two exceptionally cold winters back to back (1977 and 1978) brought this problem sharply to public attention. Newspapers and magazines were studded with statements from well-known experts describing the compelling evidence for a deteriorating climate and frightening descriptions based on reconstructions of the last ice age. Judging by the conditions that prevailed 18,000 years ago, New York City, Berlin, Edinburgh, and all the land north of these sites would be buried under a sheet of ice several hundred yards thick. So much water would be bound up in the glaciers that ocean levels would drop, baring continental shelves. Less water in the oceans would mean less evaporation and decreased rainfall. The lands that were free of ice would be subjected to prolonged drought. The deserts would grow, the forests shrink, and much of the land would consist of semi-arid loess—a yellowish-brown soil dotted with scrubby tufts of grass. Unimpeded winds would blow with relentless fury across the barren ice fields and the desiccated land. In the mid- and high latitudes, winters would be long, summers short, and the differences between day and night temperatures would be much greater than we have today.

— • —

All of these predictions seemed incredible to the members of the general public who had been accustomed to the benign climate of the 1930s and 1940s. Scientists wondered what had happened to the greenhouse effect. After careful study the suggestion was made that any warming caused by carbon dioxide was being overwhelmed by the cooling effect of more particulates in the atmosphere. These man-made clouds prevented some of the solar radiation from reaching the Earth—just as volcanic eruptions have been known to reduce the amount of incident solar energy and cause a year or two of cooler weather around the world.

Climatologists calculated that the lower atmosphere had become 2 percent less transparent since 1940, and that man-made pollution was regularly loading the atmosphere with somewhere between 8 and 24 million tons of particulates.

The largest share of this pollution came from industrial activity, a factor that was growing rapidly.

Even in nonindustrialized regions of the world, man's activities had been contributing to the thickening pall of dust. Slash-and-burn agriculture is commonly employed in primitive farming areas, burdening the air with vast quantities of smoke and ash. Under the pressure of increasing population, semi-arid lands are being overgrazed. When marginal lands are used so intensively, a slight change in precipitation patterns can turn them into bare stretches of dust and sand. A study made by the United Nations in the 1970s showed that the deserts were growing all over the world and that 6.7 percent of the Earth's surface had become man-made desert. The loose, unprotected soil created by these processes was gathered by winds and distributed throughout the troposphere, blocking out the sunlight. In some areas for weeks at a time the sky was sand-colored instead of blue and the worst dust storms brought darkness at noon.

According to several authorities writing in the 1970s the average amount of dust present in the atmosphere at that time was comparable to moderate volcanic activity and could have reduced the world's mean temperatures by almost 1 degree, thus accounting for the cooling of the Northern Hemisphere since 1940.

But then, at the very end of the 1970s, another reversal occurred in the Earth's temperatures. They began to rise, and this trend has continued throughout the 1980s, producing the warmest decade of the century. In 1987 and 1988, two unusually hot and dry summers back to back have sparked widespread concern about the future of the Earth's climate. Fears just the opposite of those expressed a dozen years earlier are receiving attention. Newspapers, magazines, and television news programs run stories carrying apocalyptic statements about the effect of a runaway greenhouse condition on Planet Earth. The media seek out proponents of a heat death on Earth and put descriptions of this general type on page one:

Scientists using the most advanced computer models predict a warming of the Earth at an ever accelerating rate. Carbon dioxide levels will double by the middle of the next century and the climate will be hotter than it has been at any time in the last million years. The glaciers on Greenland and Antarctica will begin to melt and the water will flow into the sea. Rising ocean waters will inundate many of the heavily populated coastal regions and the farmlands along the wide river deltas. The streets of New York, London, and Buenos Aires will be flooded and sickness will spread in the warm, fetid waters. As the oceans rise higher, the coastal cities and their satellite suburbs will be abandoned along with the rich alluvial plains of the Mississippi, the rice islands of Bangladesh, and the green fields of Ireland. In the low countries of Holland and Belgium, dikes will no longer hold back the seas. While the coastal areas are flooded the interior of the great continents like North America and Russia will suffer prolonged drought. The disruption to agriculture will be so severe that there will not be enough food to support the growing populations of the Earth. Famine will be widespread and wars will be fought for possession of the diminishing supplies. The ultimate consequences could be second only to a global nuclear war.

Reading these statements, I am reminded of a cautionary tale told by Mark Twain in *Life on the Mississippi:*

In the space of one hundred and seventy-six years the Lower Mississippi has shortened itself two hundred and forty-two miles. That is an average of a trifle over a mile and a third per year. Therefore, any calm person, who is not blind or idiotic, can see that in the old Oölitic Silurian Period, just a million years ago next November, the Lower Mississippi was upward of one million three hundred thousand miles long, and stuck out over the Gulf of Mexico like a fishing rod. And by the same token any person can see that seven hundred and forty-two years from now the Lower Mississippi will be only a mile and three-quarters long. . . . There is something fascinating about science. One gets such wholesale returns of conjecture out of such trifling investment of fact.

This story dramatizes the danger of taking a trend that has manifested itself for a relatively short period of time and projecting it at the same rate far into the future. Theoretically, all scientists are aware of this fallacy of extrapolation, but when faced with an important controversial issue they do not always remain aloof and impartial. Like the rest of us, they may become convinced of one point of view and argue passionately for it, passing lightly over the many unknowns and qualifying conditions. In the company of their peers this behavior is not a problem because, in the normal course of science, these strong positions are balanced out by the community of scientists who bring their varying points of view and areas of expertise to bear on the question. As a whole, then, science moves ahead cautiously, step by step. But in speaking to the general public strong advocates may make statements that appear to be based on a higher degree of certainty than is justified by the facts.

Stephen Schneider, one of our most prominent climatologists, had these words of caution for his colleagues: ". . . scientists should minimize arguing publicly for unverified—or often unverifiable—model results, particularly when dealing with lay audiences, unless they make the effort to explain clearly and frequently the wide range of uncertainties accompanying the results."

Most climatologists are stating today that two hot summers—even one warm decade—do not represent a significant trend. The droughts of 1987 and 1988 may be ac-

counted for by El Niño. The warmer decade may be just
another minor fluctuation such as we have seen occurring
throughout time, fluctuations caused perhaps by the rise
and fall of volcanic action or changes in solar activity.

On the other hand, there is widespread concern in the
scientific community about the steadily increasing concentration of greenhouse gases in the Earth's atmosphere. Although the precise consequences of these changes are
uncertain, they could be large enough to destroy the stabil-
ity of the Earth's climate system and by the time accurate
predictions can be made it may be too late to restore the
balance. It is this sense of urgency that runs in a compelling
way through all the pronouncements of even the most cautious scientists in this field.

An issue like the future of the Earth's climate is unusual in
that it affects the whole world in very important ways. Under pressure spurred by public concern, policy makers are
demanding more definitive answers, and several teams of
modelers armed with high-powered computers are working
at high speed to provide the answers.

Scientists make a climate model by setting up mathemati-
cal equations that express the basic principles governing the
atmosphere. These are fed into a computer that calculates
their effect on variables such as temperature and rainfall
under differing conditions. Any set of conditions can be
inserted—for example, increased concentration of greenhouse gases. And the computer works out a prediction
based on this scenario.

But even the most sophisticated modern computers cannot handle the vast number of calculations needed to simu-
late the atmosphere's complexity. In fact, the climatologists
themselves have not quantified all the factors. So it is not
surprising that the results vary from model to model. The
more detailed the information requested, the more unreli-
able the prediction. If we ask how doubling the concentra-

tion of carbon dioxide will affect the rainfall in North Dakota as compared with Florida, the answer produced by one model may be the reverse of the answer obtained by another. Society is really demanding more certainty from this science than is possible, given the present state of our knowledge.

Under these circumstances, the results from the computer models should be considered the best available estimates at the present time, and even these are significant only in the broad, general areas in which they all agree. Interpreted in this way, the results from the computer models tell us that if the carbon dioxide levels rose to about 600 parts per million (assuming that all changes due to such factors as volcanic action, solar activity, and orbital variations are relatively small), a global temperature increase of somewhere between 2.7 and 10 degrees Fahrenheit would occur.

Even the lowest of these figures (2 to 3 degrees) imply significant changes that would be reflected in wind and precipitation patterns. Since polar regions would be expected to warm more than low and midlatitudes there would be a decrease in the temperature difference between equator and poles. This difference controls the movement of air currents around the planet, so the flow pattern would slow down; the position of the jet winds and the band of westerlies would shift. The effect would be approximately comparable to the change from winter to summer in the world today.

There was one period in Earth's history that can serve as a rough model of the kind of changes we could expect if temperatures were several degrees warmer than they are now. The time between 4,000 and 8,000 years ago is known as the Altithermal. A great deal of very careful research has been done to reconstruct the conditions that existed at that time in various parts of the globe. By studying the fossils of ancient vegetation preserved in lake beds and peat bogs, the nature of the plant growth has been determined. Sediments have revealed ancient lake levels and erosion patterns. These facts have been pieced together to produce a picture of the climate during the Altithermal.

The areas that are now subtropical deserts—the Sahara, the Australian Outback, the southwest desert of the United States—were apparently wetter than they are now. On the other hand, midlatitude regions of North America seem to have been drier. These conclusions are borne out by historical records. We know that large cities and forests once flourished in North Africa. The land between the Tigris and Euphrates Rivers, the site of one of the earliest civilizations, was more fertile than it is now. In North America the prairie seems to have extended farther east into what is now the Corn Belt, indicating less rainfall. On the whole, however, the world appears to have been wetter than it is today and this conclusion is compatible with climate models. A warmer atmosphere can evaporate more water from the oceans. Therefore, rainfall would increase; the monsoons that bring rain to India and the Sahel would be more intense.

In using this historical example, however, we should be aware that the Altithermal is not a perfect analog for warming caused by greenhouse effect because the reasons for climate change in these two cases are different. The warmth of the Altithermal was probably due to increased solar radiation in the Northern Hemisphere summer. The planet was nearer the Sun in its orbit during the summer months than it is now, and there was also half a degree greater obliquity in the Earth's axis. While the summers were hotter, the winters may have been a little colder. But imperfect as the analog is, it gives us a sense of the degree of adjustment that might be necessary if summer temperatures increased by several degrees. The change seems to be within a range to which human beings could adapt, although there would be some shifts in favorable agricultural regions, causing important economical and political dislocations.

On the other hand, if we consider the consequences of the largest figures in the range predicted by computer models (9

or 10 degrees) we would have to anticipate an entirely different scenario, one that could approach some of the apocalyptic descriptions that have seized the imaginations of the
media and the general public. These temperature changes
are as large as the swings between glacial and interglacial
periods. They could create an environment that would be
much less habitable for man—where oceans would flood
large portions of the land and changing rainfall patterns
would desiccate the breadbaskets of the Earth. One probable
consequence of a much hotter planet would be an increase in
violent storms. Hurricanes are spawned in warm seas;
therefore in this case we might witness the fulfillment of the
prophecy set forth in the Old Testament (Hos. 8:7): "they
have sown the wind and they shall reap the whirlwind."

— • —

To put all this in perspective let us examine, then, what
Mark Twain called "the trifling investment of fact." Two
reconstructions of the Earth's surface temperature have been
made at the Goddard Institute for Space Studies and the
Climatic Research Unit. These studies suggest that a global
warming of .9 degree Fahrenheit has occurred during the
past 100 years. Another study covering the years 1901 to
1984 by the National Oceanic and Atmospheric Administration and involving 1,219 stations in the United States indicates that there has been *no* long-term upward trend in
average temperatures in the forty-eight contiguous states.

In the meantime carbon dioxide levels have increased approximately seventy parts per million in the last 100 years. A
new technology has just recently made available historic
data on levels before that time. We now have detailed information about the air our ancestors breathed and, much farther back, the air breathed by Bronze Age and even Stone
Age man.

This lost history of the atmosphere has been preserved in a deep-frozen state in the great glaciers of the Earth. The invention of techniques for drilling long cores out of glacial ice and analyzing the contents has revealed this history. Rigs similar to those used for deep-sea drilling of the ocean bottom are transported in sections by helicopter onto the ice sheets of Greenland and Antarctica, and the glaciers in Switzerland. With this equipment cylindrical cores many hundreds of feet—even a mile—long have been lifted from the ice. Fossil air is preserved in bubbles that were trapped as layer after layer of snow fell and did not melt in summer. The bottom layers were compacted, melted, and re-crystallized. By the time a snow fall is buried 200 feet deep it has become solid ice, except for tiny bubbles of air that still lie embedded in it. The polar air is in general very clean, and the ice provides a perfect container.

The release of this ancient air in a pure form, uncontaminated with air in the lab, is a process that has been worked out by atmospheric physicists at the Institute for the Exact Sciences in Bern, Switzerland. A core is cut into cubes about the size of small sugar lumps. A cube is dropped into a vacuum chamber, which is promptly sealed and all the air is pumped out. Then steel needles are pressed through a grid in the chamber, crushing the ice to bits. The air that escapes from the ice is sucked into a tube where a beam of infrared light is used to measure the concentration of the various gases present in the sample.

Each long ice core contains a compressed history of the natural and man-made changes that have affected the atmosphere. The major volcanic eruptions such as Tambora and Krakatoa and the great eruption of Santorini are clearly marked by traces of sulfuric acid. The rise and fall of the ice ages is marked by relative changes in oxygen-16 and oxygen-18. The youngest layers at the very top of the core tell the story of mankind's impact on the atmosphere. For example, the amount of lead in recent Greenland snow is 200 times the amount in prehistoric times.

Concentrations of carbon dioxide have been studied with special interest because of the concern about the greenhouse effect. They have been determined to an accuracy of a few parts per million, and the results are very striking. The level of carbon dioxide in the air has changed sharply over the last 160,000 years—a time that encompasses two major ice ages. As each ice age descended, the carbon dioxide levels dropped dramatically to between 190 and 200 parts per million. As the ice ages broke up, the levels rose to between 260 and 280 parts per million.

These figures tell us that swings in carbon dioxide levels have accompanied drastic changes in the Earth's ice cover. But it is not clear what precipitated the changes. Obviously they were not caused by human intervention. The changes in carbon dioxide may have been the *result* of temperature changes rather than the *cause*. For example, as the oceans cooled off they may have been able to hold more gases such as carbon dioxide in solution.

The only conclusions that can be drawn from these facts at the present time is that important changes in carbon dioxide levels have occurred in the past and these have been associated with glacial buildup and retreat. The level of 280 parts per million was characteristic of the period just before the beginning of the Industrial Age, and levels today are higher than any that have been discovered in the glacial ice.

On the other hand, the global temperature increase of .9 degree Fahrenheit is only half the amount that should have occurred during the last century according to the most recent climate models. This is one of the reasons why many climatologists do not yet proclaim that this observed temperature increase was caused beyond doubt by the greenhouse effect. They point out that there are a number of countervailing factors that could have influenced the data.

For example, volcanic activity was more prevalent at the beginning of this century than in subsequent years. The time from 1880 until 1912 was a period of frequent eruptions; the dust and other particulates thrown up into the stratosphere were sufficient to cool the global temperatures during those years by as much as 1 degree Fahrenheit. Furthermore, a number of other variables are not yet properly accounted for in the models (solar activity, orbital variations, and cloud cover). There are also feed-back processes that are not fully understood and that could affect the predictions.

The Earth has a considerable ability to maintain its equilibrium in the face of changing conditions. Almost like a living thing, it appears to have a self-regulating system, actively manipulating the environment and maintaining conditions most favorable for its own existence. British scientist James Lovelock has drawn attention to this phenomenon, calling it Gaia after the Greek name for the Earth Goddess. Looking at the whole sweep of geologic history, Lovelock says, the oceans and the climate have remained remarkably stable within the very narrow range that is favorable for life. Living things existed on the planet even 3.5 billion years ago and, subsequently, there has never been a time when the planet was incapable of supporting life. Even though there have been great extinctions and ice epochs when many species died out, life burgeoned again and spread exuberantly across the face of the Earth.

Historically, these major adjustments have occurred on a very long-time scale, measured in millions of years. Just as with a living organism, small adjustments are made continuously, but a change too drastic and too rapid causes the disruption of the whole system.

Several factors have resulted in a more stable climate than might have been expected from the rapidly increasing use of fossil fuels. It is an interesting fact, for example, that the amount of carbon dioxide accumulating in the atmosphere year by year is slightly less than half as much as the amount poured into it annually. The other half is stored somewhere else in the Earth system—in the oceans and the biosphere. The effectiveness of these storage mechanisms as carbon dioxide levels increase are large unknown factors in the climate modeling equations.

The oceans are capable of holding immense amounts of carbon dioxide in solution—perhaps sixty times as much as the air can hold. But the amount of storage that can take place within the time span relevant to human affairs is limited. When carbon dioxide levels in the atmosphere rise, the oceans absorb some of this excess, coming back into equilibrium with the atmosphere. However as the waters warm, the amount that can be held in solution decreases, so higher concentrations in the air result in both storing and releasing of this gas as temperatures rise. To calculate the net effect is a complicated problem involving ocean circulation patterns and the turnover time of surface waters. It is just the top 300 feet of ocean that participates directly in the interaction with the air, and this layer is replaced on the average about once a century. To involve the deep ocean water would require a time span of almost 1,000 years.

There is another storage process that involves the oceans as well as the biosphere and the atmosphere. Carbon dioxide in the air dissolves in rainwater, creating a slightly acidic precipitation. When this rain falls on silicate rocks (which are very common on the Earth's surface) it weathers and erodes them. Reacting chemically with the silicates, it produces calcium and carbonate ions that are washed down by rivers and streams into the sea. There they are used by the vast population of tiny marine organisms to construct shells of calcium carbonate that eventually fall to the ocean bottom and are gradually compacted into limestone and other carbonate rocks. Since warm surface temperatures in the ocean increase the amount of evaporation and, therefore, the amount of rain, the weathering process is accelerated by rising temperatures and more carbon is stored in shells. But this storage mechanism, too, has a slow response time—a matter of many millennia.

The white cliffs of Dover are a beautiful example of this process. They are almost pure calcium carbonate, composed of trillions of skeletons of microscopic zooplankton. Here in these cliffs lies an enormous storehouse of carbon dioxide locked up in carbonates. Each of the tiny creatures whose shells are compacted here lived for only a few weeks or months, but their forms have been preserved for as long as 70 million years in this soft-textured white chalk. Sometime in the past a major tectonic change in the earth's crust lifted them high above the depths where they had been laid down. Now they are gradually being eroded by rain and returned to the sea.

— • —

Another very large store of carbon is sequestered in the organic compounds of all the living matter that covers the Earth. By photosynthesis, green plants remove about 100 billion tons of carbon dioxide from the atmosphere (14 percent of its total carbon content) every year. The amount varies with the season. In spring and summer when the plants are passing through an active growing phase, the carbon dioxide concentration in the air increases by about 3 percent and decreases by this amount in the fall and winter. Thus we see that levels of carbon dioxide respond very quickly to the changing demands of plant life.

A portion of the carbon stored in plants is taken up by animal life and further sequestered. Through respiration both animals and plants expel some of the carbon immediately as carbon dioxide. Eventually all living things die;

their bodies decay and one of the products of this process is carbon dioxide that flows back into the atmosphere.

The relative rates involved in this cyclical exchange are critical in determining the net storage or release of carbon dioxide. As we have noted earlier, when the concentration of this gas increases in the atmosphere and if sufficient amounts of sunshine and water are available, photosynthesis increases, producing more luxuriant growth of vegetation and removing some of the excess carbon dioxide from the air. On the other hand when temperatures increase, the rate of decay and respiration also increase and more carbon dioxide flows back into the atmosphere.

This carbon cycle operates on a shorter time scale than the cycle involving storage of carbon in the sea because plant life responds very quickly to changes of carbon dioxide levels. Increase in the rate of decay takes place a little more slowly. So if all other factors remained constant, the vegetation on Earth might act as a regulator, restoring favorable levels of carbon dioxide as small fluctuations in it occurred, and this would take place on a continuous basis before temperatures on the Earth's surface had increased sufficiently to affect the rate of decay.

But unfortunately, other factors are not remaining constant. Mankind is speeding the death and destruction of plant life, thus accelerating the processes that convert carbon stored by photosynthesis back into carbon dioxide. The cutting and burning of forests has been going on for centuries as men have cleared the land for agriculture, used wood for fuel, and built their houses and boats. Recently with the explosive growth of human population and technology, the rate of deforestation has increased dramatically. Slash-and-burn agriculture is taking place in many poor areas of the world, and now the greatest reservoir of plant life on Earth is being put to the torch.

The Amazon basin—an area almost as large as the conti-

nental United States—contains about one-third of the
Earth's remaining tropical forest. This is the richest and
most diversified biota anywhere on the planet. One hectare
(two and one-half acres) of the rain forest contains more
plant species than all of Europe. Spurred by economic policies that favor the clearing of forest for agricultural land and
for huge development schemes, this forest treasure is being
rapidly destroyed.

Satellite pictures of the Amazon basin show hungry
tongues of flame licking across the lush expanse of green
forest and great plumes of smoke drifting up to disperse
throughout the atmosphere. One photograph shows 6,000
man-made fires burning in the forest on a single day. These
pictures document the fact that an area nearly as large as
Kansas was burned in one year alone (1987).

Ironically, the land that is cleared is not good for farming.
The soil is poor because it has been leached year after year
by heavy rainfall. But the burning of the forest is taking
place not just to create more farmland. Huge industrial
projects have been undertaken—mining and smelting oper-
ations. Roads have been built to accelerate the building of
these projects, and many of these have been financed by
multilateral development banks like the World Bank and
Inter-American Development Bank—with money provided
by the industrialized nations.

Concern about the greenhouse effect has presented the
world with a sensitive problem. Is it possible to protect
natural resources that are important for the ecological balance of the whole Earth and still help nations like Brazil
participate in the advantages of the modern industrial age?

Several hopeful signs have appeared in recent months.
Local Brazilian groups have been formed, promoting the
preservation of the forest and using its products in ways that
provide an income for the people who live there. These
groups, working with the development banks and the gov-

ernment, have set aside large forest areas for sustainable harvesting of rubber, Brazil and resin nuts, and special plants selected from the forest. Twelve such "extractive reserves," totalling roughly the area of New Jersey, have been created.

But whenever a change in policy occurs, some people stand to lose while others gain, and frustrated greed can promote violence. One of the leaders of the movement to create these reserves was threatened by a coalition of men who expected to benefit from the continued destruction of the Amazon forest. Francisco Mendes told the Brazilian papers: "If a messenger from the sky came down and guaranteed me that my death would strengthen our struggle, it would be worth it. But experience teaches us the contrary. It's not with big funerals and motions of support that we are going to save the Amazon. I want to live." Three weeks later he was shot and killed at his home on the edge of the rain forest in northwestern Brazil. Outrage at the action has echoed around the civilized world.

In the meantime organizations such as the World Bank and the Inter-American Development Bank are becoming more aware of the environmental consequences of the development projects they fund. In early 1989, the Brazilian President José Sarney announced a new program to reduce the destruction of the Amazon forest. He pledged action to prohibit the exporting of lumber, to suspend government subsidies for farming in the Amazon, and to study possible measures to halt the "predatory development . . . destroying our flora and fauna." If such reforms are implemented the whole world will have reason to be grateful.

It is estimated that deforestation adds somewhere between .5 and 2.5 billion tons of carbon to the atmosphere every year. This reduction of green vegetation on the Earth's surface results in a reduction of the annual uptake by photosynthesis. Such changes, of course, increase the carbon dioxide level in the atmosphere.

Furthermore, carbon dioxide is not the only villain in this scenario. There are at least three other gases that contribute significantly to the greenhouse effect—methane, nitrogen oxides, and chlorofluorocarbons. Methane is perhaps the most important of these three, although it is a gas that is produced by nature to a greater extent than by man's activities. There is much less methane in the atmosphere than carbon dioxide, but molecule for molecule it is twenty times more effective as a greenhouse gas.

Methane is a byproduct of the processes of respiration and decay in environments such as rice paddies, swamps, and bogs where oxygen is not freely available. It is present in natural gas deposits and in coal mines. As soil warms, the rates of decay and respiration increase; so, theoretically, methane levels would also increase with rising global temperatures, thus acting as a positive feed-back mechanism that would accelerate a warming trend.

One surprising source of methane is present in almost every barnyard. An average cow expels at least 100 gallons of methane a day. In fact, many animals such as buffalos, sheep, deer, and elephants produce large quantities of this gas as cellulose is broken down in the rumen (the first section of their stomachs) into digestible cud. A herd of dairy cows placidly chewing their cuds in a pasture, are quietly making their contribution to the greenhouse effect. It has been proposed that we might give cows an antibiotic to inhibit the bacteria that produce methane and decrease atmospheric levels of this greenhouse gas. The suggestion is typical of the quick-fix solutions that are apt to do more harm than good.

The other two gases that contribute to the greenhouse effect—nitrogen oxides (from cars, and coal burning) and chlorofluorocarbons (from spray cans, refrigerators, and air conditioners)—will be considered in later chapters.

With all these contributing factors—from cars to spray cans and livestock, from power plants to the poor farmers of

Brazil—it is a wonder that the temperatures of the Earth have not risen more dramatically since the beginning of this century. The fact is that they have increased globally just half the amount predicted by the climate models. How have we been spared a swift descent into hothouse Earth?

One possible explanation is the increased load of particulates in the atmosphere. The air is less transparent today than it was at the beginning of this century. This turbidity is especially visible at dawn when the Sun's light is veiled by the long trip through air loaded with tiny droplets of sulfuric acid, with ash, and with the other effluents of industry. Even in the countryside far from the major sources of pollution, ground smog is a common phenomenon on summer mornings.

Measurements taken at places like Mauna Loa cannot record the extent of this increasing opacity because, unlike carbon dioxide, the particulates have a relatively short residence time in the atmosphere. They are deposited on the ground or washed out in rain and snow. But they are continually replaced in increasing quantities.

The air over the ocean is relatively free of this burden. In fact the average particulate concentration over land in the industrial parts of the world is ten times as much as it is over the oceans. This is one of the reasons why islands are such lovely places. Tahiti, Hawaii, Bora Bora, Bali, Sri Lanka, Madeira—the very names conjure up delightful images of sun-filled days bright with many flowers, of star-studded nights and the full Moon's glittering wake on the restless seas. Islands are little remnants of the way the many land surfaces of the Earth used to be before mankind began altering the atmosphere.

Although the increasing turbidity of our air is degrading our environment in many ways, it may be one of the factors that has kept the temperatures on Earth moderate in spite of increasing concentrations of greenhouse gases. Another factor may be increased cloud cover because, of course, as more particulates exist in the air, more clouds form. Most experts agree that cloud cover would also increase as the Earth warms. However, the effect of this increase on the Earth's climate is the weakest link in the theoretical model. The most recent information we have on this subject is the study mentioned earlier (see page 39) of satellite measurements taken and analyzed by a team of scientists, headed by V. Ramanathan at the University of Chicago.

The results of this research show that the clouds surrounding the planet in 1985 reduced the absorption of incoming solar radiation by a considerably larger amount than they reduced infrared losses to space. The net reduction in the radiant heating of the Earth was surprisingly large— several times greater than the warming predicted from the greenhouse effect if carbon dioxide concentration is doubled. Based on these figures an increase in cloud cover of about 25 percent could be sufficient to counteract the predicted warming. However the scientists point out that increased cloud cover might work the other way if the clouds that formed in a warmer climate were placed differently in relation to the land masses of the Earth and at different altitudes in the atmosphere. The net effect of the increase might enhance instead of counteract the greenhouse phenomenon. In fact, most of the current models assume that changes in cloud formations under a strengthening greenhouse trend would amplify the warming. The numbers involved in this uncertainty are very large, so this is an extremely important unanswered question that could completely change the prediction for the next century.

For the sake of argument, let us suppose that the clouds would be favorably placed and that they would counteract the warming trend. This seems to offer a happy escape from hothouse Earth. But it may not be a completely beneficial one. If this happens we will be trading a hot future for a murky and cloudier one. How will the Earth's biosphere respond to darker days? Reduced photosynthesis might alter the delicate balance and precipitate changes that would adversely affect all life on Earth.

— · —

In the panic generated by the recent publicity about the greenhouse effect, it is not surprising that some suggestions have been made for interventions by man. One such proposal is the deliberate spreading of dust in the stratosphere to artificially create more clouds and turn more sunlight back into space. Another suggestion is to cover the Earth's oceans with white Styrofoam chips, which would reflect light more efficiently than ocean water. What frightening examples of the dangerous hubris of mankind!

This beautiful planet can best be protected by reducing the impact of human activities, and by a careful and thoughtful approach to problems tempered with a little humility, because there is so much that we still do not understand.

6

— . —

A Bitter Wind Blowing

Earth, air, stone, and water are continuously altering one another. Bad Branch Falls, Kentucky. (Photograph by Earl Dotter.)

The wind that blows out of the north and sweeps down the Rhine Valley toward the city of Cologne is a bitter wind. Tinged with a sharp, acrid odor, it darkens the sky with a brown haze. It echoes through the narrow streets of the old city and envelops the majestic spires of the Gothic cathedral. In this delicate stone fretwork, it seeks out every surface modeled and polished by artists' hands six centuries ago. The blanket of corrosive air does not spare the most sacred symbols of faith. The angels lost their wings and the prophets looked like victims of leprosy with their noses eaten away. In spite of heroic efforts to save this architectural treasure, its beautiful face has been permanently scarred with this life-threatening illness, and the disease is spreading in epidemic proportions across the continents of the Earth. It has attacked the marbles of the Parthenon and the flying buttresses of Notre Dame. The Statue of Liberty has needed extensive repair and the Lincoln Memorial has suffered irreparable damage.

The winds that bring destruction to the Cathedral of Cologne carry their poisoned air farther south down the Rhine River to the Black Forest, where trees are dying in a plague so serious it is known in Germany as *Waldsterben* (death of the forest). The effect spreads southeasterly into Bavaria, where in 1982 it was reported that 150,000 acres of forest were composed largely of dead trees and another 8 million were showing early signs of damage. Across the border in Czechoslovakia's Ore Mountains more than 270,000 acres have died and efforts to reforest the land have been failing.

When the winds turn and blow from the south across the Ruhr Valley, the pollution is swept north into the Scandinavian peninsula, a land of deep fjords and spectacular scenery, where many districts are still largely untouched by man's activities and thousands of small lakes are set like blue sapphires in dark evergreen forests. These lakes that used to abound with salmon and trout were favorite sites for sportsfishermen. But in the 1950s fewer fish were found there, and inspectors identified the problem as unusual acidity of the water. Monitoring over the next few decades revealed a steadily increasing problem. By 1983, fish life had been destroyed in 5,000 lakes in Sweden and 1,500 lakes in Norway. Salmon and trout were the first to go, and restocking proved to be useless. Many of these lakes are now totally devoid of fish.

— . —

Because the threat of acid rain is an issue that has just recently been recognized, research was needed to establish the relationship between the damaging effects that had been observed in many places and their possible causes. As in a felony case, the perpetrators must be tracked down and their guilt firmly established. The case is now proven beyond reasonable doubt for two of the charges—destruction of stone and metal structures and killing of aquatic life. The evidence on the third charge—the murder of forests—is circumstantial, but very convincing.

By chemical analysis it has been shown that acid precipitation results from the sulfur and nitrogen oxides poured into the atmosphere by the burning of fossil fuels—especially high-sulfur coal. Water, in the form of rain or snow, is provided by nature in a very pure state. When water molecules evaporate from the lakes, rivers, and oceans around the world, they are separated from any adulterating compounds that may be present, drawn up into the troposphere, and recombined as cloud droplets and snowflakes. But on their way down through the atmosphere they become altered by reactions with the various chemicals they encounter.

Carbon dioxide, as we know, is one of the gases that is always present in the atmosphere and it combines with water, forming carbonic acid. This reaction results in a mildly acidic rain that erodes the silicate rocks of the continents, but on a time scale that is long compared to the lifetime of a man. Limestone and marble, which are favorite building materials, are carbonate rocks and are also affected by carbonic acid, but very, very slowly. Acids produced by sulfur and nitrogen oxides reacting with water attack the carbonate stones much more vigorously and erode them rapidly.

Acidity and alkalinity are measured by pH numbers from one to fourteen. A pH of seven is neutral; lower numbers measure increasing acidity and higher numbers indicate stronger alkalinity. The scale is logarithmic, so a change of the pH by one means a ten-fold difference. Normally rain has a pH of 5.6, but rainfalls with a pH as low as two have been measured in some areas. This is comparable to the acidity of lemon juice, 4,000 times that of normal rain.

The reaction cycle that creates acid rain is triggered by sunlight striking a molecule of ozone, which is present to some extent even in the lower atmosphere. In this reaction a molecule of oxygen gas is formed plus a highly reactive single oxygen atom, which then combines with a water molecule to form two *hydroxyl radicals*. Composed of one hydrogen atom and one oxygen atom, these radicals are very active forms of matter. Although they are extremely scarce in the atmosphere (only one in a trillion molecules) the supply is inexhaustible because they act like catalysts. They trigger a reaction and then at the end are regenerated again. These hydroxyl radicals transform nitrogen dioxide into nitric acid, and they initiate the reactions that transform sulfur dioxide into sulfuric acid.

Sulfur and nitrogen compounds are released into the atmosphere from a number of sources. Some of these are natural such as vulcanism and the activity of soil bacteria. The very hot effluent that erupts from vents on the ocean bottom is heavily laden with sulfur. But human activities have added enormously to the natural load of sulfur and nitrogen in the atmosphere.

From the study of ice cores scientists have been able to measure the concentration of these compounds in Earth's atmosphere at various times in its history. Analysis of a recent core from south Greenland, covering the years 1869 to 1984, showed that sulfate concentrations have tripled since the turn of the century and nitrate concentrations have doubled since about 1955.

The fact that acidic precipitation is a regional phenomenon provided clues to the major sources of these polluting gases. The distribution was first noted in England more than a century ago. Sulfate concentrations in rain were observed to be greater in densely populated locations and especially near coal-burning plants. Acidity of water standing in bog ponds was strongest near industrial areas. In the 1950s, scientists who monitored the composition of rain at stations in Sweden noticed that episodes of regionally high concentrations of sulfuric and nitric acid were related to the direction of the wind. When it came from the south or southwest, transporting air from southern England, the Benelux countries, and central Europe the levels of acidity were high compared with air coming directly from the North Atlantic. This information caused widespread concern in Europe, and the problem of acid rain became a major environmental issue there.

Awareness of this issue was later in coming to North America, although measurements of precipitation taken as early as the 1950s showed some increases in acidity in the northeast United States. This phenomenon was blamed at first on natural sources, but later analysis identified a trend of increasing acidity of precipitation in these states relative to the rest of the country. At the same time naturalists were beginning to hear reports that many lakes in New England

and Canada were losing their fish populations. By 1970, aquatic life had been virtually destroyed in hundreds of lakes in the Adirondacks and Ontario. Matching this information with airflow patterns, scientists suggested that much of the acidity in these areas originated in the industrial Midwest and was carried in on the prevailing winds.

There were, however, a number of troublesome facts that were difficult to explain on this basis. Some lakes that were downwind of the heaviest industrial areas did not seem to be affected while others were. Some of those that were adversely impacted had not shown these signs immediately. Years might elapse between the first significant exposure and the death of their aquatic organisms. A study of the chemistry of the various lake waters and the surrounding terrain provided an explanation for these differences.

Some lakes are set in land rich in limestone; the soils are naturally alkaline and can neutralize the acid that falls on them, much as Alka-Seltzer or a Rolaid tablet relieve an acid stomach. The rocks on the watershed are eroded by rain; so the water that flows into the lake carries with it carbonate ions that react with the acids, producing neutral compounds. As a result the pH of the water does not fall significantly and the life of the organisms is not endangered.

Over a period of time, however, even lakes that were originally alkaline may become susceptible if they are regularly exposed to strong levels of acid precipitation. The buffering action of their soils is gradually reduced. In an experimental acidification of a small lake in Canada, when 70 percent of the alkalinity was depleted, pH levels dropped sharply and fell detectably below normal.

This is an example of a principle we have encountered throughout nature. If the challenge to natural systems is kept within reasonable limits, it may be counteracted by processes that restore the equilibrium of the system. But if the challenge is too great or occurs too rapidly, it overrides the forces that are acting to restore balance and the system is destroyed. This is called a *threshold* phenomenon; the stimulus must reach a certain level before its presence is perceptible.

There are many lakes—some of our most beautiful ones—whose waters have little or no buffering capacity. Any input of acid rain will show up in stronger acid content immediately. These lakes lie in regions of predominantly igneous rock or sandy soil. Igneous rocks, such as granite and quartz, are resistant to weathering and do not have the proper chemistry to neutralize acid. Many of the lakes in Maine, Vermont, New Hampshire, and upper New York State are very susceptible for this reason.

In Canada, very ancient igneous rocks are widely exposed in the eastern two-thirds of the country and the margins of Lake Superior. This formation is called the Canadian Shield. Younger rock and sediments that once overlaid the Shield have been eroded away, leaving rocks like granite and basalt exposed on the surface. Lakes that occur in these regions are not protected by any buffering action, and many of them are already showing signs of damage.

Lake George in Ontario's Killarney Provincial Park has long been a source of inspiration for artists. It is a scene of wild beauty, set in glittering quartzite rocks and surrounded by mountains. It teemed with fish as late as the 1950s. Now its waters are still crystal clear, but they support no life.

— • —

The fish die because low pH interferes with the salt balance that freshwater species need to maintain in their body tissues and blood plasma. More moderate degrees of acidity kill the fish eggs and the fry. The critical levels vary for different species, but they are most frequently experienced in the spring when the snow melts, dumping an unusual amount of acid precipitation into the lakes and rivers. These events usually coincide with spawning time.

There is another factor that poses a threat to human health as well as to aquatic organisms. Metals that are present in soils—aluminum, manganese, zinc, nickel, lead, mercury, and cadmium—are "mobilized" by the action of acidic rain. The metals are usually present in the soil in compounds that are relatively harmless, but the reaction with certain acids releases metal ions that can be ingested by biological organisms. In living bodies, the metal ions can accumulate in concentrations that are toxic to human beings. The Swedish government warned the public a few years ago against eating livers or kidneys of moose because these organs had concentrated cadmium that had been ingested by moose browsing on vegetation in acidified waters. Some livers contained levels of cadmium that would be fatal to human beings. Mercury and lead are also extremely dangerous. Lead can be leached from pipes carrying drinking water in regions of acid rainfall. The presence of these metals in the food chain opens a Pandora's box of poisons.

An acidified lake loses many species in addition to its fish. Types of algae, crustacea, mollusks, and insects are also wiped out. Stoneflies and mayflies disappear. Since these insects normally feed on dead leaves that are floating on the lake, their absence results in an unusual accumulation of decaying matter on the lake bottom and disturbs the balance of the ecosystem.

The severity of these effects appears to be directly related to the degree of acidity and can be reversed temporarily by heavy applications of crushed limestone. But the treatment is not a cure; it is only palliative, and after a few years the acidity returns. All this information has convinced scientists that acidity is causing these lakes to die.

— • —

The impact of acid rain on vegetation has been more difficult to document because there are more variables to consider in a terrestrial ecosystem than in an aquatic system. However, laboratory experiments suggest that acidic conditions can reduce the resistance of plants to infection and invasion by insects. Simulated acid rain has been shown to result in reduced growth and yield in some garden vegetables and certain types of grain. At Brookhaven National Laboratory plots of soybeans were exposed to rain with different levels of acidity. Rain with a pH of four caused a 2.6 percent reduction in seed yield. Such small changes are difficult to observe in the field because the natural variability from year to year is greater than 2 percent. But if this reduction is real it would represent a significant economic loss.

The most dramatic damage to vegetation has occurred in the evergreen forests at high elevations. Measurements of acidity taken at various altitudes in the atmosphere have shown that maximum acidity occurs in fog and near the base of cloud formations. High altitude forests, immersed in clouds, are directly exposed to the unusually strong acidity that exists at the cloud base.

One particularly interesting long-term study has been conducted at Camel's Hump, a 4,100-foot peak in Vermont's Green Mountains. In 1965, Hubert Vogelmann, Chairman of Botany at the University of Vermont, was interested in cataloging the forest communities on this mountain. Even near the summit "the trees were luxuriant," he recalled. But by the late 1970s he saw a steady decline in the population of red spruce. There was no evidence of disease, and the decline was not related to years of drought. By 1980, when almost half of the red spruce above 3,000 feet had died, Vogelmann and his colleagues accepted an offer made in the *New York Times* by American Electric Power Company (the largest producer of sulfur dioxide east of the Mississippi River) to fund a study on the effects and causes of acid rain.

Throughout the following decade regular surveys of Camel's Hump revealed progressive declines in balsam fir as well as red spruce near the peaks. The birch community farther down was also affected, and there was a surprising decline in the mature sugar maple forest near the base of the mountain. Small sections drilled from living trees indicated steadily declining growth rates that correlated closely with historically documented increase in acid precipitation. And chemical analysis showed a tripling of aluminum in the rings since 1950. Aluminum, you remember, is one of the metals mobilized by acid rain.

Vogelmann believed that their studies firmly established a cause and effect relationship between the death of the forest and the pollution carried on the prevailing winds from the industrial Midwest. But when the reports were reviewed by American Electric Power Company executives, they failed to see a connection between these two.

"Our reading of the reports," said their environmental engineer, "would indicate that there is some sort of stress in the area, but in terms of the cause of the stress we can't tell. Nothing in the material we've looked at ties down acid rain as the factor."

The hypothesis that death is due to stress is probably correct, but one cause of the stress may be acid rain. There are many natural factors that adversely affect forest communities: disease, high winds, ice storms, and shortages of water and essential nutrients. Isolated stresses can usually be handled successfully; the biological system has reserves that can be called upon to restore health. But when stresses act simultaneously, the total effect may be fatal. A tree might be weakened, for example, by a severe winter. Then it is defenseless against the continuing stress of acid rain.

Investigators have identified a number of ways in which acid precipitation could create disadvantageous conditions in forest communities. The soil loses nutrients when vital elements have been leached away by acid rain. Aluminum released from soil and taken up by the trees may compete with calcium for binding sites on fine roots, slowing the growth of the tree. Strong acidity and concentrations of aluminum can reduce populations of bacteria that release nutrients from decaying vegetation and injure the symbiotic fungi that live on the roots of conifers and contribute to the health of the tree.

The needles of conifers, frequently bathed in acidic rain and cloud droplets, may become leached of important nutrients faster than they can be replaced. Furthermore, the process known as cold hardening, which normally takes place in the fall and prepares the tree for winter, may be interrupted by false signals. The process is triggered by a decrease in nitrogen produced by microorganisms in the ground. But nitrogen compounds deposited by acid rain may override this signal and leave the tree vulnerable to damage caused by ice forming in needle tissue. None of these possibilities has been definitely proved, but laboratory experiments now underway may provide answers within the next few years.

In the meantime efforts continue to identify the contribution of each source to the acid precipitation in the affected areas. The problem is a difficult one because cause and effect are separated from each other in both time and space. In situations where industrial plants producing sulfur and nitrogen oxides are located in forest land and the effluent is emitted from relatively low stacks, it is easy to see the clear correlation between the pollution source and dead trees in the immediate vicinity. But recently very tall stacks have been built in an effort to meet local air standards. Tall stacks may relieve regional pollution problems but they export the pollution to other lands downwind. Industry spokesmen do not admit that this is a serious problem. The emissions from tall stacks, they claim, "are dissipated high in the atmosphere, dispersed over a wide area, and come down finally in harmless traces."

The assumption that emissions from tall stacks are dissipated does not take into consideration the complex movement of air masses in the troposphere. As we know, the air does not just move up and away, resulting in uniform dilution. Its movement is constrained by various temperature and pressure conditions, frequently causing a layered effect that holds strong concentrations of chemicals close to ground level. Because the Earth's surface cools off more rapidly than the air at night, the atmosphere in contact with the Earth cools off first, and this heavier layer of stagnant air, often laden with pollution, remains near the surface. Under the proper atmospheric pressure conditions these inversion layers may persist for many days and nights.

Over and above these local phenomena, we must not forget that the whole troposphere is a relatively shallow portion of the atmosphere, and it acts like a giant inversion layer. Chemicals added to it are not carried away into space. They never truly disappear; they come down, borne in cloud droplets, in rain, and snow, and are deposited again on the Earth's surface. The radioactivity released into the air at Chernobyl, for example, was not dissipated harmlessly. It showed up in alarming concentrations thousands of miles away.

In an attempt to track down individual sources responsible for acid rain, scientists are using sophisticated computer models to plot spatial distributions of the chemicals. The amount of sulfur or nitrogen gas released at each plant can be measured directly at the source. By using tracer gases the movement of effluent plumes can be tracked. An inert gas that remains capable of detection for thousands of miles is released at ground level near the site to be studied, and the transport of this gas is monitored by a network of stations.

From these studies it has been found that sulfur dioxide

and nitrogen oxides and the acids produced by them remain in the atmosphere for several days and are carried distances of 250 to 1,000 miles. Sulfuric acid condenses to form microscopic droplets that are ideal condensation nuclei and thus are important precursors in cloud formation. These droplets also contribute to the haziness of the atmosphere in the summertime throughout the northeastern United States. Satellite photographs have shown that these smoggy conditions can extend over large areas and last for many days.

Most of the sulfur dioxide produced by man's activities comes from plants that burn high-sulfur coal without using any technology to remove the sulfur gases before the effluent is emitted. Power plants using this type of fuel are the most important sources. In 1970, new regulations in the United States required that all new plants of this kind be fitted with sulfur abatement technology, but plants built before that time were not retrofitted. The coal that is mined in Ohio and West Virginia is high in sulfur content, and this coal is used in a number of large, old plants on the Ohio River. It is not surprising, therefore, that Ohio is the state with the largest sulfur dioxide emissions. This state emits two-and-one-half times as much per square mile per year as the average emission in the entire northeastern United States (the region bounded by North Carolina and Tennessee on the south and the Mississippi River on the west). These emissions are substantially more concentrated than the standards presently proposed for sulfur dioxide. A plan to take 10 million tons of sulfur dioxide out of the emissions every year targets nine states that would be required to accomplish this reduction, and Ohio is one of those states. But industry executives and government officials protest that the plan places too great a burden on this state, and powerful lobbies are attempting to block the proposed legislation.

A National Academy of Sciences report released in June

1983 concluded that an increase or decrease in sulfur dioxide emission into the atmosphere would result directly in an equal increase or decrease in acid precipitation. If each state in the heavily industrial region of the Midwest reduced sulfur dioxide emissions by 50 percent, the level of acidity in precipitation downwind of that region would similarly be reduced by 50 percent.

This simple linear relationship was confirmed by a recent study in the western states. Most of the sulfur dioxide emitted in the region between the Sierra Nevada and the Rocky Mountains originates from copper smelters in Arizona, New Mexico, Utah, and Nevada. Copper production and, therefore, smelter emissions fluctuated widely between 1980 and 1983 because of labor problems and an uncertain market. A peak occurred in 1981, followed by a sharp drop in 1982, and a slight increase in 1983. Graphs of the acidity of rain and snow falling in the Rockies as far away as Colorado, Wyoming, and Idaho during that period of time show an almost exact correspondence with the pattern of rise and fall of copper production. "These results provide powerful evidence—for the East as well as the West—" said Michael Oppenheimer, who directed this study, "that if we reduce sulfur dioxide going up, we reduce the acid rain coming down." This demonstration was possible because of special conditions in the western states where a few very important isolated sources were all responding to the same market fluctuations. Such a correlation could not have been demonstrated in the East where the emissions are relatively constant and come from more diverse sources.

The largest single source of sulfur dioxide emission in North America was the International Nickel Company's copper–nickel smelting plant in Sudbury, Ontario. It is equipped with a stack nearly a quarter of a mile high, the tallest in the world. Before 1983, 5,000 tons of sulfur dioxide were emitted every day from this stack and carried east on the prevailing winds across evergreen forests and the hundreds of little lakes nestled on the Canadian Shield. When the Canadian public became aware of the threat to their lakes and forests, they demanded action and the government responded. The Province of Ontario ordered a reduction in sulfur emissions at Sudbury to 1,950 tons a day by 1983. These standards have been met, and further reductions are planned to reduce the sulfur emissions to the lowest feasible level. Until 1980, the Canadian provinces had the right to set the emission standards for sulfur and nitrogen oxides. But soon it was recognized that the pollution emptied into the air from sources—particularly those with very tall stacks—is not just a local problem. The law was amended, and the Canadian Parliament now has the right to set the standards for pollution that crosses provincial boundaries.

Like the carbon dioxide problem, acid rain is a phenomenon that affects large portions of the world. The corrosive wind is no respecter of national frontiers. In many cases the regions that suffer most lie outside of the countries where the pollution is produced. Canada now emits a relatively small amount of acid-creating gases, but her eastern provinces lie downwind of the industrial midwestern United States and are suffering damage far greater than the amount they inflict on others. Scandinavia is defenseless against the acid winds that blow from the industrial districts of England and central Europe. The economic dislocations are very important; some nations are absorbing much more than their share of the cost, while others take advantage of lower costs in the production of power and manufactured goods.

Although sulfur dioxide is the major acid-producing gas released by man's activities, we must not forget the role of the nitrogen oxides. They are responsible for about 25 percent of the problem. These gases are produced by high-temperature burning processes—in power plants, smelters, steel mills, and automobile engines. In fact all types of transportation including diesel trucks, aircraft, and farm machinery produce nitrogen oxides. In the United States, tailpipe emissions of nitrogen oxides have been significantly reduced in response to clean air laws—about a 76 percent decrease in gas-powered new cars. But older cars and all vehicles equipped with diesel engines continue to emit large quantities of nitrogen oxides into the atmosphere. Tighter regulations controlling the release of these gases have been proposed, but the auto industry fears that the public would not be willing to pay for the increased cost of the changes that would be required.

— · —

A startling new discovery brings into focus the fact that acid rain is not just a regional phenomenon in industrial countries. It is an international problem. In June 1989, scientists reported that acid rain had been found over the virgin rain forests of Central Africa. Low pH levels comparable to those in Europe and North America were found throughout the year in an area stretching from the Congo basin in the heart of Africa to the west coast and extending out over the Atlantic Ocean. The researchers, who had come from West Germany and France to study this problem, felt certain that the condition in the rain forests was caused by man-made fires that rage for months at a time across thousands of miles of African savannahs. The farmers and herdsmen set the fires to clear the land of low-growing shrubs and stimulate growth of the grasses.

This burning increases the atmosphere's load of fine particulates as well as nitrogen oxides and significant amounts of carbon dioxide, which can enter into reactions that form ozone at low altitudes in the troposphere. Northeast winds move the tainted air into the tall, dense forests of equatorial

Africa. From the Ivory Coast to the Congo and Zaire, acid fog and mist envelop the broad leaves of the trees and rain delivers the acid to the soil and root structure. Some of the acids produced by savannah burning are familiar ones— carbonic acid from carbon dioxide and nitric acid from nitrogen oxides. Other less familiar ones are acetic acid and formic acid, which are both compounds of carbon, hydrogen, and oxygen. Ozone in surprising quantities is also present in the air just above the rain forests, and ozone is the trigger that starts the reactions, converting certain gases to acid. It is strange to think that this lush green canopy of foliage, far from human population centers, is bathed in a man-made brew of highly reactive chemicals.

No one knows yet just how serious this pollution is to the rain forests of Africa, but the question must be raised and the results monitored with care. We cannot afford to lose another large proportion of the forest cover of Earth. The jungles of the Amazon have been decimated, those of Central Africa are threatened, and the evergreen stands of Germany and Bavaria, of Scandinavia, and the Adirondacks are dying.

The green mantle of trees that still clothes many of the mountainsides and river valleys of Earth is a purifier of our atmosphere and an important regulator of our climate. Through the chemistry of photosynthesis, trees absorb carbon dioxide from the atmosphere and store the energy of sunlight in organic compounds. They shade us from summer heat and break the biting winds of winter. They protect the watershed and manage the rainfall. The deep network of root fibers holds the soil in place and the many layers of vegetation on the forest floor absorb the rain and return it at a more leisurely rate to the surrounding countryside. In mountainous regions, the leaves and pine needles continuously comb the clouds and fog, collecting tiny cloud droplets, gathering them into raindrops that spill on the ground and keep the soil moist even in very dry weather. The leaves, still damp, are exposed to every breeze that blows, and the air receives the moisture back again as it removes heat from the leaf. So pure water returns to the atmosphere, leaving pollution behind. Thus, evaporation continuously cools and sweetens the air.

A walk in one of the mature forests of the Earth is an experience in tranquility. The light is softened as in the nave of a great cathedral; the air is hushed and fragrant. The majestic columns of the tall tree trunks reach toward the sky and arch into a lacy fretwork of green. The eye and the spirit are swept upward following these soaring lines. It is easy to imagine that the architects of the great Gothic cathedrals like Cologne and Amiens were inspired by a walk in one of the virgin forests of medieval Europe and they sought to re-create this spiritual serenity in stone.

If we allow the forests of the Earth to be destroyed, we will have lost an irreplaceable source of joy and inspiration. And we will have destroyed an essential element in nature's finely balanced ecology. Greenhouse gases will accumulate more rapidly in the atmosphere. Lands lacerated by erosion will put greater pressure on the remaining fertile areas of the planet. Destructive agricultural practices will be harder to control and the process will feed upon itself.

— • —

The problems of greenhouse effect and acid precipitation are closely related to each other. They are both caused by the burning of fossil fuels and vegetation. They are both international in scope, although the causes and possible effects are unevenly distributed around the world. The effects of acid rain cross national frontiers, and greenhouse warming would impact the whole Earth.

In several ways, however, these problems differ sharply.

The damage caused by acid rain is clearly visible, and it is more easily controlled than the problems that result from the buildup of carbon dioxide. The residence time of sulfur and nitrogen oxides in the atmosphere is a matter of days, but many of the processes that remove carbon dioxide from the air are related to slow geologic changes—the turnover time of oceans and the storage of carbon compounds in the Earth's rocks. These cycles move in long rhythms—at least a thousand years, perhaps many millions.

If the release of acid-producing gases were cut in half and kept at that level, then the destruction of buildings and vegetation and aquatic life would be reduced by a half or even more because there is a threshold effect with acid rain. Natural systems can be exposed to a certain amount of stress without being permanently damaged. But when the stress exceeds these thresholds, the damage becomes irreversible.

Fortunately, practical technologies exist today for controlling the emissions that cause acid rain. Sulfur compounds can be removed in several ways: before the fossil fuel is burned, during the burning process, or it can be washed out of the gases before they are emitted. Nitrogen oxides have been more difficult to control until quite recently. But a very promising new technology has just been developed at Argonne National Laboratory. It is said to be capable of removing more than 70 percent of nitrogen oxides produced in coal-burning plants, and it would represent a relatively small additional expense when combined with scrubbers that remove sulfur dioxide from stack gases.

Of course, these corrective measures will cost money. They will increase the price of electricity and of the goods produced in large coal-burning plants. Estimates for the increased price of electricity vary widely, from 2 to 50 percent, depending on who is making the estimate— environmentalists or representatives of the electric industry.

A more realistic figure is provided by the experience of the Tennessee Valley Authority who, in response to a lawsuit filed by the Natural Resources Defense Council, undertook a 50 percent reduction in sulfur dioxide emissions from their massive coal-fired generating system. This program was completed in 1983, and the utility rates charged their customers increased by only 6 to 8 percent in the most expensive year. However, the rest of the utility industry has dragged its feet about accepting the need for cleaning up their emissions. Their position has been that we do not know enough about acid rain to warrant spending sufficient money to control it.

In answer to this position, Stephen Schwartz, writing in *Science*, made a comment that is unusual in this serious scientific journal. When industry spokesmen asked the question, do we know enough to regulate? Schwartz replied with another question, "Do we know enough to *emit*?"

Every day that we as a society allow more acid-producing gases to be poured from the hundreds of aging plants lining our industrial centers, we are making a policy decision on this issue. We continue to pollute our atmosphere, add poisonous chemicals to our food-chain, cause the death of our lakes, threaten our great forests, and destroy our artistic heritage in order to avoid paying any increase in the price of goods or power.

The cost of any product or service is a measure of its value to society, and industry believes that it is responding to the demands of society in keeping the price of luxury as low as possible. People want bigger cars that will go faster and are loaded with more extras. So don't add cost for a benefit that does not show! Cheap electricity can be equated with more air conditioning, more microwave ovens, more color televisions, more computer games. Many of the people who buy these things do not recognize the hidden cost that has not been paid. This debt is accumulating at compound interest.

I think that industry has misjudged the American people. If they understood what is happening to this wonderful environment that we inherited here on Earth they would be prepared to pay a reasonable price to preserve it, to pass it down unspoiled to the next generation.

This beautiful planet where flowers brighten the brown crust of the continents and cloud shadows make moving patterns of light and darkness across the seas—this paradise has been 4 billion years in the making. The atmosphere has evolved in a close, symbiotic relationship with the developing membrane of sensitive living things. It has gradually changed to meet life's needs and has been changed by it. The result is an environment uniquely suited for the nurturing of life. As far as we know, nothing like it exists anywhere else in the solar system—perhaps in the whole universe.

As we have explored space we have discovered an astonishing diversity of atmospheres among our sister planets. From the concentrated greenhouse gases and acid clouds that obscure the hot surface of Venus to the cold, poisonous atmosphere that lies with crushing weight on the surface of Saturn, we have surveyed a wide gamut of uninviting possibilities. Of all the satellites of the Sun, only Earth is endowed with the perfect combination of warmth, moisture, and the essential ingredients for life.

In some ways we have been blessed by good fortune. Our planet orbits at a favorable distance from the Sun; so water can exist in a liquid state here. It is the right size to hold an adequate but not too heavy atmosphere. However, the remarkable nature of our atmosphere cannot be ascribed to chance alone. It was created by eons of coevolution of life and its environment. Now the fine balance that has been achieved is threatened.

The cosmonauts and astronauts who have been observing our planet from space have seen that changes have taken place just in the last few years. A Russian cosmonaut remarked that the atmosphere over Lake Baikal is now as polluted as the atmosphere over Europe. And an American astronaut observed that fifteen years ago he could take much clearer pictures of the industrial centers than he can today.

The changes mankind is making in this delicate blue halo of air are happening too rapidly to be accommodated by the slow processes of evolutionary change. The time scale for nature to make such adjustments is measured in millions of years. We cannot afford to wait a million years; even a century or a decade is too much. We must act now to restore this balance if we care about ourselves and the other ephemeral beings that share with us this brief instant of Earth time—the mayflies that dimple the still surface of a pond on a summer day, the whippoorwills whose clear calls echo in the woods at dusk, and the sweet-faced violets that carpet the forest floor in spring.

7

— · —

A Hole in the Ozone Umbrella

Glacier in Antarctica. (Photograph by Sharon Chester.)

By a strange twist of fate, Antarctica—the place most isolated from human activities—is the site of the most dramatic sign of man's impact on Earth's atmosphere. Over this continent, a hole develops every year in the thin veil of ozone—the "umbrella" that shields all life on Earth from the destructive ultraviolet light of the Sun. Ironically, too, the conditions that lead to the formation of this hole develop during the long, terrible polar winter, when there is no sunshine for several months.

To witness the slow retreat of the Sun is an experience almost impossible to imagine for those of us who live in temperate latitudes. But Admiral Byrd has given us a compelling firsthand impression of the coming of the polar night:

Each day more light drains from the sky. The storm-blue bulge of darkness pushing out from the South Pole is now nearly overhead at noon. The sun rose this morning [April 12] at about 9:30, but never really left the horizon. Huge and red and solemn, it rolled like a wheel along the Barrier edge for about two and a half hours, when the sunrise met the sunset at noon. For another two and a half hours it rolled along the horizon, gradually sinking past it until nothing was left but a blood-red incandescence. The whole effect was something like that witnessed during an eclipse. An unearthly twilight spread over the Barrier, lit by flames thrown up as from a vast pit, and the snow flamed with liquid color. . . .

The coming of the polar night is not the spectacular rush that some imagine it to be. The day is not abruptly walled off; the night does not drop suddenly. Rather the effect is a gradual accumulation, like that of an infinitely prolonged tide. Each day the darkness, which is the tide, washes in a little farther and stays a little longer; each time the day, which is a beach, contracts a little more, until at last it is covered. The onlooker is not conscious of haste. On the contrary, he is sensible of something of incalculable importance being accomplished with timeless patience. The going of the day is a gradual process, modulated by the intervention of twilight. You look up, and it is gone. But not completely. Long after the horizon has interposed itself, the sun continues to cast up a pale and dwindling imitation of the day. You can trace its progress by the glow thrown up as it makes its round just below the horizon.

With the Sun's warmth withdrawn, intense cold settles in. The Antarctic winter is the most severe weather experienced anywhere on Earth. Now a special type of cloud begins to form, much higher than most clouds—above the troposphere in the thin, warmer air of the stratosphere. Like mother-of-pearl clouds they catch the Sun's rays after it has retreated over the edge of the world. This silvery reflection is one of the few bright features in the sky during the long darkness of the Antarctic winter and even this light is heavily veiled by clouds that are massed in the troposphere.

In this forbidding scene, where winds blow with unrelenting violence month after month, very few life forms remain on the surface of the continent. Small groups of men cluster in scientific stations where they are warmed and protected from the weather. Outside in the bitter wind, lichen still clings to the rocks; and, on the ice shelves bordering the land, the male Emperor Penguins engage in a strange annual ritual. Assembled in large groups, numbering hundreds or even a thousand, they huddle together, facing toward the center of the circle with their backs to the wind. Each penguin holds on his feet a single egg laid by his mate who has now left for the warmer environment of the sea. The incubation period is sixty-two days and during this time the males do not eat except for an occasional mouthful or two of fresh snow. They can move only a few inches at a time or the egg might roll away and be broken. The survival of the species depends upon this extraordinary communal behavior. A single penguin alone on the ice would not live through the winter. Together they survive by breaking the wind for each other, sharing their body warmth, and incubating the egg that will carry life forward into the future.

In the little scientific centers, which several nations have established in Antarctica, the men spend many of the long winter hours studying samples of the life forms that exist in the sea and taking measurements of the polar atmosphere—temperature, pressure, wind speed, and the chemistry of the air.

For twenty-five years the British Weather Station at Halley Bay has been monitoring the concentration of ozone in the column of air over Antarctica, and this information has been relayed back to England. At the same time, measurements have been taken by instruments aboard the American satellite Nimbus 7. These readings have been collected at NASA's Laboratory for Atmospheres at the Goddard Space Flight Center.

Since 1977 the British team had been getting anomalous readings in the Antarctic spring, but these were so far off normal that the technicians thought they must represent instrumental error and did not include them in their reports. The same false assumption was programmed into NASA's computer. It was instructed to throw out readings that deviated by more than an arbitrary amount below the norm. The fact that this happened every October when Nimbus passed over Antarctica was not noticed as a significant fact until the leader of the British team, Joseph Farman, became suspicious of the very low readings that had been measured at his station every September and October. Even though these findings had not been corroborated by the data from NASA, Farman decided to go public with his information. His paper, published in May 1985 in *Nature*, caused considerable excitement and concern in the international community of atmospheric scientists.

According to the satellite data (which had not been reported, but had been saved) the concentrations of ozone dropped every year in late September, eventually decreasing to a fraction of the normal value (as low as one half in 1987) and remaining at low levels throughout October. Then each November it climbed back up to the average amount and increased steadily with the onset of the polar summer. This phenomenon began about 1977 and the depletion has continued to become more severe since that time. Although there were fluctuations year by year, the hole had in general grown deeper as time passed. By the late 1980s, this patch of thin ozone covered an area the size of the continental United States. Now there is some evidence that it is beginning to affect parts of the other southern continents.

About 1974, scientists first raised the possibility that the chemicals known as chlorofluorocarbons (CFCs) might destroy ozone in the stratosphere. These chemicals have a number of special characteristics that make them very useful. They are extremely inert molecules that do not react with other substances in the lower atmosphere. They are nontoxic and nonirritating, so they were thought to be remarkably safe to use as propellants for sprays. They are also very efficient as heat-exchange fluids in air-conditioning and refrigeration systems, and they are useful as blowing agents in the manufacture of many plastic foam consumer products. Although large amounts of the gases were being expelled into the atmosphere from these devices (approximately 1.7 billion pounds a year in 1973), they were believed to be free of any side effects.

However, it occurred to a chemistry professor, Sherwood Rowland at the University of California at Irvine, that this inertness might be altered by exposure to very intense ultraviolet light—just the kind of light, in fact, that is absorbed by ozone in the stratosphere. This hypothesis was checked in the laboratory and found to be correct. When struck by high-energy ultraviolet radiation, CFC breaks down, yielding, among other products, one or more chlorine atoms. Each of these atoms, acting as a catalyst, transforms one single atom of oxygen and one molecule of ozone into two molecules of normal oxygen. In the process the chlorine atom is released and is ready to perform its catalytic function again.

The inertness, which was an environmental advantage at Earth level, becomes a great disadvantage when CFCs start their wandering trip up through the atmosphere. Borne on the erratic updrafts that normally mix the air throughout the troposphere, a molecule may take a decade to reach the tropopause, and most molecules are so unstable or reactive that long before they reach this level they have changed their identity or have been washed out in rain. But CFC molecules make the trip intact and arrive eventually at the upper edge of the troposphere.

Above this point, the temperature inversion stops the general upward movement. The CFC molecule, however, is in no hurry; its stable structure gives it a long life expectancy. Eventually the circulation patterns of the stratosphere carry it up to heights of approximately twenty miles, where it meets radiation of sufficient strength to decompose it and liberate the free chlorine atoms.

The same atmospheric conditions that delay the diffusion of CFCs into the stratosphere cause a lengthy residence time for the chlorine atom in the ozone layer. It remains there either in the free state or as chlorine monoxide (which participates in the catalytic destruction of ozone) for one or even two centuries. Because of the protracted time spans involved in the transport of CFCs to the stratosphere and their long residence there, the effect does not respond quickly to a reduction in emissions. Chemists estimated in 1976 that if the release of these chemicals were halted immediately the ozone reduction would continue for at least a hundred years.

But the release was not halted in the late 1970s. Although the United States did ban its use in most types of spray cans, the use worldwide continued to increase. During the next decade measurements (including those taken at the British Weather Station and on Nimbus 7) began to reveal slight decreases in the concentration of the ozone layer throughout most of the world. But no one was prepared for the news of the dramatic depletion taking place over Antarctica every austral spring.

Since the discovery of the hole, chemists and meteorologists have been working hard to identify the causes of ozone destruction as well as the characteristics that make Antarctica unique in its seasonal response. Although CFCs had been implicated, it was also known that certain oxides of nitrogen in the presence of ultraviolet light could act as catalysts, initiating reactions that destroy ozone, turning it back into normal oxygen. Most nitrogen oxides created at ground level are not stable enough to reach the stratosphere by the usual slow processes of natural air movement. Nuclear weapon testing, however, provides an efficient transportation system, delivering the products of high temperature reactions in air straight into the stratosphere. The heat of the explosion causes an updraft as powerful as a hurricane directly above the site of detonation. The rapidly rising air current acts as a chimney sucking all the products of reaction high into the atmosphere. Fortunately, nuclear testing in the atmosphere has been restricted, so this source of stratospheric nitrogen oxides has not become an important factor. But scientists point out that a nuclear war could so deplete the ozone layer that any life still remaining on Earth would be destroyed by the ultraviolet rays of the Sun.

There is one other way in which nitrogen oxides can reach the stratosphere. And this process depends not on the enor-

mous force of a nuclear explosion but on the action of tiny
bacteria living in the soil on Earth's crust. The *denitrifying
bacteria* help to process decaying organic matter. They undo
the work of the nitrogen-fixing bacteria, converting nitrates
back into gaseous nitrogen. Some nitrous oxide is also produced, a chemical that is popularly known as laughing gas.
This is an unusually stable oxide of nitrogen and it can make
the long trip up through the atmosphere to the ozone layer.

In recent years, nitrogen fertilizers have been added to
farmland to encourage rapid growth of agricultural crops.
Any excess not used by the plants remains in the soil and is
broken down by the denitrifying bacteria, adding to the
quantity of nitrous oxide liberated at the Earth's surface.
This new source of the oxide is a subject of increasing concern because it also acts as a greenhouse gas contributing to
global warming and because the use of fertilizers worldwide
will undoubtedly escalate as populations grow and more
efficient use of farmland is demanded.

It turns out, however, that the presence of nitrous oxide in
the stratosphere is a minor factor compared to the free chlorine atoms derived from the CFCs. In fact, nitrous oxide
enters into reactions with chlorine atoms in combination
with oxygen and water, and the end result of these reactions
is chlorine nitrate, a relatively inactive compound. Thus, two
of the principal catalysts for the destruction of ozone are
sequestered together. These compounds do eventually
break down if they encounter an energetic photon of light or
react with other chemicals, but for a while at least they form
an inactive reservoir, making the catalysts unavailable to
attack ozone.

— • —

Because of its extremely cold winter climate, Antarctica is
the only place in the world where considerable cloud cover
regularly forms so high up in the atmosphere, actually in the

ozone layer. These clouds consist of small ice crystals, and other molecules that float in this rarefied air tend to collect on the surface of these crystals. Chlorine nitrate molecules are among those attracted to the cloud crystals. Many chemical reactions that occur slowly when molecules are moving freely in a gas take place more rapidly on the surfaces of particles. No one knows exactly what reactions occur on these ice crystals throughout the long, dark "days" of winter. But when the Sun returns in the spring and the atmosphere warms, the ice crystals melt and ultraviolet light shines on the tiny droplets. Then the compounds break down and chlorine atoms are liberated, free to pursue their usual activity of ozone destruction.

Wind circulation patterns in the south polar region also contribute to the isolation there of an air mass that is not regularly mixed with other air from the Southern Hemisphere. In the stratosphere, air circulates from tropical latitudes (where the most ozone is formed) toward the polar regions, carrying with it ozone-rich air. In the Northern Hemisphere, this air circulation usually carries all the way to the pole, and ozone levels there reach quite high levels. But, in the Southern Hemisphere during the coldest months, the circulation pattern stops at about 60 degrees, impeded by the polar vortex, a tightly twisted whirlpool of winds. In the late spring, the polar vortex dissipates, allowing a rapid influx of air from lower latitudes. Then the concentration of ozone is restored.

The breakup of the polar vortex, however, allows ozone-depleted air to drift away from the polar region and dilute the ozone layer over the southern portions of New Zealand, Australia, South Africa, and South America. Evidence of this phenomenon was reported in scientific journals in 1989. It occurs in December, the beginning of the summer months in these regions, when the Sun is high and the people are outdoors, beginning to enjoy the beaches. This thinning of the protective layer is sufficient to cause about 14 percent more ultraviolet to reach ground level. So on the brightest days, people who live in these parts of the world may be advised to stay indoors to protect themselves from the sunshine.

The North Pole does not experience the extremely low temperatures that are a regular winter phenomenon in the Antarctic, so ice clouds in the stratosphere are not as extensive and the particles do not remain frozen as long into the spring. Although, in the wintertime, a polar vortex exists there as well, it is much weaker and breaks up earlier. These milder conditions in the Arctic account for the fact that the dramatic losses in ozone observed over Antarctica are not duplicated at the North Pole. However, researchers who have studied the conditions in the Arctic winter are concerned that considerable destruction can also take place there.

In 1988 and 1989, scientific teams flew into the ozone hole in both Antarctica and its potential counterpart in the Arctic. A modified version of the U-2 spy plane was turned into a flying laboratory, equipped with computerized instruments to take continuous measurements of the chemistry of the atmosphere. This plane carries just one man and is capable of flying very high. It cruises comfortably at altitudes over 60,000 feet, right into the heart of the ozone layer. Another plane is also used, a DC-8 which cannot fly as high but carries a number of scientific observers and a much heavier load of equipment for making remote observations.

The information collected from these flights has verified the important role of the chlorine atom in the destruction of ozone. Surprisingly high levels of chlorine monoxide are present in the ozone hole, and the inactive reservoirs of chlorine nitrate are depleted during the spring months in Antarctica. In the Arctic, too, chlorine monoxide levels are elevated, although not as high as in the Antarctic hole. The chemistry of both these portions of the stratosphere is more

perturbed than theorists had expected and now there is
concern that the integrity of the ozone layer may be more
severely threatened than we have recognized. Measurements taken in September and October 1989 indicate as
severe a loss over Antarctica as occurred in 1987, the year
that previously held the record for the deepest hole. Annual
variations have occurred and have seemed to fall into a
biennial pattern throughout the last decade. According to
Susan Solomon, atmospheric scientist at the National Oceanic and Atmospheric Administration in Boulder, Colorado,
the reasons for these annual variations may depend upon the
severity of the Antarctic winter, but the reasons for this
particular weather pattern are not clear and, in fact, may
prove to be coincidental when records have been kept for a
longer period of time.

— • —

While this research into the polar regions has been taking
place, regular satellite readings taken over the Northern
Hemisphere have revealed a significant decrease in the concentration of the ozone layer between 1978 and 1985. In
latitudes 29 to 39 degrees north, the decrease has been somewhere between 5.7 and 1.7 percent. Farther north, the decrease has been slightly smaller, between 4.4 and 1.0
percent. The larger of these figures approaches real danger
levels for human beings, as greater amounts of very short-
wave ultraviolet (UV-B) reach ground level.

   One of the most serious dangers for human health is the
increased risk of skin cancer, including malignant melanoma, a rapidly spreading and frightening form of cancer.
According to studies conducted in the United States, the
prevalence of most types of skin cancer varies geograph-
ically; people living in the South have a higher incidence
than those living in the northern states, and these observa-

tions correlate with increasing exposure to sunlight and with the decreasing thickness of the ozone layer in lower latitudes. In 1974, the National Cancer Institute made a survey of the number of cases of the two most common forms of skin cancer (squamous-cell and basal-cell carcinoma) in Dallas and Minneapolis. The Dallas figures (four cases per 1,000 population) were more than twice as large as the number of cases in Minneapolis (one and a half cases per 1,000), where the ozone layer is known to be about 30 percent thicker. It is unfortunate that the lower latitudes, which receive the greatest amount of solar radiation throughout the year, are less well protected by the ozone layer.

Throughout the process of evolution, nature responded to this health hazard in the equatorial and tropical latitudes by developing protective skin pigmentation to screen out ultraviolet rays. If the ozone layer undergoes a sudden change, it is the fair-skinned people who will suffer the greatest damage. Projections vary widely, but it is possible that a 5 percent reduction in the ozone shield would cause 30,000 additional cases of skin cancer in the United States every year and 500,000 worldwide.

Although the correlation between the amount of ultraviolet irradiation and incidence of malignant melanoma is less well documented, there is evidence that the disease is linked to exposure to sunlight. Statistics from a 1989 publication by the American Cancer Society support this conclusion. And, furthermore, the incidence of this very threatening form of cancer is found to be rising rapidly in all countries—with increases of about 3.4 percent a year.

Shortwave ultraviolet light is also injurious to many other forms of life. Small organisms such as plankton in the sea are very susceptible to increased ultraviolet irradiation. Since plankton serves as the base for the aquatic food chain, all forms of life in the sea would be affected if the ozone layer were depleted. The tropical seas contain fewer of these microorganisms compared with the profusion present in the oceans of the higher latitudes.

The seas around Antarctica are abundantly supplied with plankton and the small shrimp-like crustaceans named *krill,* which feed on them. The krill in turn serve as an important link in the food chain that serves most of the higher forms of life that have been able to live so successfully in this hostile environment: the seals, the seabirds, the porpoises, and the whales, as well as the vast populations of penguins that waddle on the iceshelves and swim like fish in the dark seas. In a large penguin colony there are as many as a million birds, and each one eats about four pounds of krill a day. Thus, each community places a very heavy demand on this one food supply in the ocean surrounding the colony. Skuas and leopard seals dine on baby penguins as well as krill, and the krill are absolutely dependent on the plankton for their food. So the depletion of the plankton would endanger the whole chain of life in this polar region.

When the presence of the ozone hole was first reported in 1985, biologists began to wonder whether the increased radiation of ultraviolet light every spring constitutes a threat to this fragile but very prolific ecosystem. It was already known that large amounts of UV-B can kill organisms or render them incapable of reproduction. The most energetic rays of sunlight damage organisms by attacking the DNA molecules, which carry the blueprint for each living thing and pass this blueprint on to the next generation. When DNA is altered, the consequences are almost always detrimental. These mutations can cause cancer, sterility, and many different types of deformity. Smaller amounts of UV-B can slow down the process of photosynthesis.

Although photosynthesis occurs only in the presence of sunlight, a portion of the Sun's radiation interferes with this wonderful process on which all life depends. It proceeds at a slower rate when UV-B is present. The mechanism responsible for the inhibiting effect is not well understood, and no one knows how rapidly the decline may take place. But there is some concern that the hole in the ozone layer may cause a serious decrease in photosynthesis in the Antarctic seas.

Experiments designed to find an answer to this question are underway at America's Palmer Station. The photosynthetic activity of phytoplankton has been tested under varying degrees of ultraviolet exposure. When samples were irradiated with large doses of UV-B, 50 percent higher than the natural exposure during the most extreme depletion episode of 1987, the photosynthesis decreased dramatically (a two- to fourfold change). Under natural exposure conditions, during a more moderate depletion episode such as that which occurred in 1988, the rate of photosynthesis declined about 15 or 20 percent in the top three feet of the ocean waters. At lower depths the decline fell off exponentially.

Aquatic life has the great advantage of being able to retreat to greater depths in order to protect itself from ultraviolet light. This is how life was able to survive and proliferate in the ancient seas before the ozone layer formed (more than 700 million years ago). Almost thirty feet of water is needed to completely shield the organisms from these lethal rays. But the light needed for photosynthesis is attenuated at these depths, so photosynthesis cannot take place as efficiently. A very fine balance must be struck between enough visible light for the photosynthetic process and protection against UV-B. Of course, if the ozone layer were seriously depleted around the planet, the plant life on land would have no way of protecting itself as the aquatic life does and photosynthesis would decline. How large this effect would be is not yet known, but it would be undesirable in any degree. All the animals on Earth, including mankind, depend indirectly on photosynthesis for their food supply. To inhibit this activity would mean a reduction in crop yields everywhere in the world, and in many places even today there is not enough food to sustain the exploding populations.

Also, we must not forget that photosynthesis is one of the principal ways in which carbon dioxide is removed from the atmosphere. If the rate of photosynthesis declined, the greenhouse effect would increase—a result that we can ill afford at the present time.

Furthermore, the CFCs, which are known to be causing a large part of the ozone depletion, are very effective greenhouse gases themselves. Although they represent, at the present time, a much smaller component of the atmosphere than earthbound carbondioxide, CFCs are believed to be responsible for almost one-fourth of the total greenhouse effect today. It has been calculated that the emission of CFCs at the present rate could cause a rise in temperature of 1 to 2 degrees Fahrenheit by the end of the century. So the CFCs threaten to increase the risk of global warming while at the same time exposing all life on Earth to increased doses of dangerous radiation.

We may ask if there is any evidence that the thinning ozone layer is allowing a significant increase in UV-B radiation at ground level. In searching for answers to this question, scientists have turned up some surprising information. The annual average amounts of UV-B for eight geographical locations in the United States *decreased* between 1974 and 1985. The fact that this has happened in spite of the slightly depleted ozone layer has led to the conclusion that something in the atmosphere is reducing the amount of solar energy that reaches Earth's surface. Pollution at ground level appears to be the most likely explanation. As one scientist remarked, we seem to be replacing the ozone shield with a pollution shield.

All of the collection sites monitored in this study of UV-B radiation levels are urban centers that have grown since

1974. And urban pollution has also grown. The presence of particulates, nitrogen oxides and sulfur oxides can interfere with the transmission of ultraviolet light, and ozone is now present in the air we breathe. It is ironic that, while we are destroying ozone in the stratosphere, we are creating it at ground level. Ozone is now present in concentrations that are believed to be damaging to human beings and to many other forms of life. The same molecule that protects us when it is high up in the stratosphere can be harmful when it comes into contact with living things.

This apparent contradiction is easy to understand if we remember that ozone is an especially active form of oxygen, and this element itself was dangerous to life forms before they evolved ways of protecting themselves against its reactive nature. Throughout the whole time that life has existed on the land surfaces of Earth (until just a few decades ago), ozone was a rare constituent of the lower atmosphere. Because the UV-B rays of the Sun were intercepted by the ozone molecules in the stratosphere, the radiation that reached ground level did not have enough energy to break up the oxygen molecule and convert it into ozone. In the absence of any compelling need, therefore, the evolutionary processes did not provide protection against ozone. The enzymes that guide oxygen safely through the biological systems cannot deal effectively with this more violent form of oxygen. But as mankind has altered the chemistry of the atmosphere, ozone has become a significant constituent of the air in the lower troposphere. This ozone, unfortunately, will never reach the stratosphere and replenish the supply there because the molecules are too reactive to make this long trip intact.

Two types of chemicals acting together with sunlight catalyze the production of ozone near the Earth—nitrogen oxides and hydrocarbons. As we have seen, nitrogen oxides (principally nitric oxide) are emitted by automobiles and burning processes that take place at high temperatures. Hydrocarbons are also emitted by automobiles and by many commercial activities such as printing presses and dry-cleaning plants. When these chemicals combine, one of the products formed is nitrogen dioxide. This is a particularly unpleasant compound; it has a bitter smell and a whiskey-brown color that darkens the smog that contains this pollutant.

The third ingredient necessary to turn the chemical brew into ozone is sunlight. It is sufficiently powerful at ground level to energize the reaction that creates ozone from nitrogen dioxide. Therefore, the greatest concentrations occur on hot summer days, and they reach a maximum in early afternoon when the sunlight is the strongest. By this time, the air has been filled with the exhaust fumes from automobiles and trucks, and power plants, working at top capacity, have poured forth generous quantities of nitric oxide. If, in addition, a temperature inversion exists, the mixture is concentrated in a shallow layer. The Sun beating down upon this layer completes the stewing of the witches' brew. In the evening, when the traffic has subsided and the Sun has gone down, ozone concentrations usually drop back to lower levels. These conditions are most frequently encountered in large cities and it has now been found that almost 100 cities in the United States have dangerous ozone levels on many summer days.

In human beings and other animals, ozone affects the respiratory system, causing decreased vital capacity in the lungs, accompanied by chest pain and cough. Drowsiness, headache, nausea, and inability to concentrate have also been reported after very brief exposures. When air temperature and humidity are high, these effects are magnified; exercise and physical labor bring on the symptoms much more rapidly.

Exposure to elevated levels of ozone causes visible damage to plants. Leaves turn yellow, and this change occurs in the cells that contain chlorophyll. In this case, the rate of photosynthesis is again significantly reduced. Damage to vegetation can occur at levels that are exceeded on approximately one-third of the summer days in the midwestern and eastern parts of the United States.

In recognition of this problem, air-quality standards setting maximum permissible levels for ozone and related species were adopted in the United States. The installation of catalytic converters on cars is required, thereby reducing the amount of hydrocarbon emission. But, in spite of these measures, ozone levels have been increasing. Every year more cities are violating the air-quality standards. A study of ozone trends in a number of urban centers in California and Texas from 1973 to 1982 showed increases of several percent a year in the annual mean concentration.

In the meantime, it was discovered that ozone pollution is not just a city phenomenon. Peak concentrations often occur during mid- to late afternoons about thirty miles downwind of city centers. These facts can be explained by the presence of natural hydrocarbons, which are produced by trees and other vegetation. The emissions are enhanced during warm, daylight hours and the rate increases rapidly as temperature rises.

Since both nitrogen oxides and hydrocarbons are required to generate ozone, the control of man-made hydrocarbons alone is not sufficient to reduce ozone levels. If nitrogen oxides are present with natural hydrocarbons on hot, sunny days, the three preconditions are met. According to this theory, reduction of rural ozone levels will require control of nitrogen oxide emissions.

This is not an impossible goal. We can control emission of nitrogen oxides and thereby alleviate the acid rain problem as well. We can also outlaw the use of CFCs. They are convenient but not essential to our way of life. Reasonable substitutes have been found to take their place and are expected to be commercially available in 1990. The cost of these changes will be small compared with the price we will pay if we take no action.

If we replace the ozone shield with a pollution shield we will be living in a murky world, heavy with an acrid odor. And even this brown haze would not protect us. On summer days, our radios might issue such warnings about high ozone levels as: "Stay indoors; remain quiet; do not take any exercise." Thus our lives would be confined more and more to artificial inside places. More covered downtown areas and recreation areas would be built. And this construction would require more plaster board, more concrete, more iron and steel. So the factories that turned out these materials would fill the air with more polluting chemicals. And the inside air would have to be filtered, warmed, or cooled. The use of electricity would skyrocket and more effluents from the burning of fossil fuels would be poured into an already overburdened atmosphere. We would be caught in a downward spiral, a self-augmenting process, constantly reducing the quality of our lives.

Imagine how much poorer we would be if we could not enjoy warm, sunny days when we can picnic on the grass and explore a mountainside where wildflowers bloom. All the luxuries we might buy with the money saved on cheaper aerosol cans and lower electric rates are but tawdry toys compared with the many luxuries that nature—unmolested—bestows upon us.

— • —

I like to spend the summer months in a part of the country where the air is still fresh and sweet with the fragrance of clover and new-mown hay. I often wake before dawn and lie in bed listening to the sounds of morning, enjoying the smell of grass wet with dew, and watching the pale light softly flood across the sky. The first heralds of dawn are the mourning doves whose gentle, long-drawn-out notes terminate the stillness of the night. The call is answered once—and again. Then silence reigns for a long pause. Suddenly a score of birds take up the song, like an orchestra striking up an overture to morning. At no other time do birds sing with such enthusiasm. It seems to be an affirmation of life, an expression of joy in the returning day.

And when twilight has descended again, I watch the fireflies light their tiny lanterns over the darkening fields. They flash their bright signals only in rising flight; so the thousands together move in waves of ascending light. No choreographer could create a ballet that more perfectly expressed the joy of being alive on a beautiful summer night.

There are many places on Earth where fireflies still dance and mourning doves still greet the dawn. It is not too late to preserve them. It is not too late to cooperate with nature. If we stop the influx of damaging chemicals into the atmosphere, nature will eventually put ozone back where it belongs—in the thin, transparent layer that filters the sunlight but does not obscure our view of the Moon and the evening star and Orion riding high in the night sky. Now it is possible to reverse the damage, but every year that goes by brings us closer to a time when it will be too late.

8

— • —

The Challenge

We have found a strange footprint on the shores of the unknown. And lo! it is our own. (First footprint on the moon. Courtesy of NASA.)

Only a little while ago human beings were powerless to affect the great sweep of natural processes. Like all the other species, we lived within the sheltering care of the Earth Goddess, who in due time would put everything back to rights if things fell a bit out of tune. The changes made in the soil and the air were small, well within Her ability to restore the balance. The food and energy taken from the biosphere were replenished by the abundant fruitfulness of Earth. The thresholds had not yet been passed. But all that is changed; now we have the power to alter the environment for every living thing and to rearrange the furnishings of the planet. We can remove forests and plant new ones. We can change the course of rivers and replace the soft soil of Earth's crust with hardtop and concrete. We can sow the wind to make clouds form and rain fall. We can even transport tons of dust to scatter in the stratosphere and put a screen between ourselves and the Sun.

In the pride of these new powers, we have ignored the many inadvertent, harmful effects that have accompanied our precipitous mastery of nature. These have been accumulating year by year, and now the evidence is all around us. We find it in the stagnant brown air that blankets our cities, in the acid winds that carry death into pristine natural areas, and in the strange chemistry of the vortex winds that blow with unrelenting strength in the polar night. We have altered the composition of the delicate medium that is our home in space. We have frayed the thin fabric of the ozone umbrella and have changed the way the radiation from our star is received here on the surface of the planet.

But the ability to reason, which gave us this power to remake the Earth for our own convenience, can also be used to arrest and even to undo the destruction we have carelessly created. We are *Homo sapiens*, the thinking species; we have the ability to identify problems and plan how to control them. We have monitored the temperature of the Earth and taken detailed measurements of the chemistry of the atmosphere. We know that the symptoms of a serious disorder are there and should not be ignored. Now all we need is the will to act on this knowledge.

— • —

The three major threats considered in these pages—the warming of the Earth, the increasing acidity of precipitation, and the destruction of the ozone layer—can all be controlled in ways that are well understood today. The effects of the latter two are clearly apparent and can be corrected with existing technology. We can ban the use of CFCs in the next few years, and although there will be a time lag in restoring the ozone layer, it will eventually be reconstituted. CFCs can be replaced with other less-destructive chemicals, and the cost will be relatively minor. Sulfur and nitrogen oxides can be removed from stack gases and better automotive design can reduce the outflow of nitrogen oxides. Since these gases do not have a long residence time in the atmosphere, the effect of this change would be felt quite rapidly in reducing the acidity of rain.

The greenhouse effect is by far the most difficult of the three problems. Because the most important greenhouse gas—carbon dioxide—comes primarily from the burning of fossil fuels, an energy source on which the whole world depends, total prevention of these emissions would involve fundamental changes that might not be acceptable to society. Very large quantities are involved; about 6 billion tons of carbon dioxide are added to the atmosphere every year. Although large amounts are also removed—dissolved in the oceans, washed out by rain and snow, and taken up by vegetation in photosynthesis—almost 3 billion tons of extra carbon dioxide are retained in the air every year.

But considering the present state of our knowledge, drastic measures that would disturb the economy do not seem to be justified. Although there is no doubt that we are altering the Earth's atmosphere in ways that could affect global temperature, there is still an element of uncertainty in the timing and degree of this change. The reasoned response in this situation is to reduce the emissions of greenhouse gases in ways that would improve our environment, even if the present prognosis of climatic change proves to be pessimistic. Fortunately, there are several options which, taken together, can considerably reduce these emissions without serious disruption to society and at the same time would contribute to the control of acid rain, ground-level pollution, and the depletion of the ozone layer. These three problems can best be solved in unison.

The banning of CFCs will not only halt the destruction of the ozone layer, but it will reduce any warming trend as the remaining CFC molecules are gradually eliminated by natural processes.

The removal of greenhouse gases could also be accelerated by the planting of many new trees. It has been estimated that 100 million trees would remove 18 million tons of carbon dioxide each year. To plant enough trees to take all the additional annual emission of this gas from the air would require a space about half the area of the United States and would cost somewhere between 200 and 400 billion dollars. These figures are only rough estimates, but they give a sense of the magnitude of the problem. Although its cost is very large, it is not an impossible option. We spend that kind of money on defense every year. A one-time reforestation project would give us relief from global warming for twenty or thirty years, during the most active growth period of the trees. This is a possibility that could serve as an emergency measure, but the more realistic option for the present is a more modest tree-planting program in conjunction with a number of other measures.

The United States Environmental Protection Agency estimates that 400 million new trees could be planted in urban areas in this country. Placed strategically, they would beautify the cities and shade the buildings, cutting down on the demand for electricity to air-condition in the summertime. The savings thus realized would help to finance the expense of the planting.

Government programs could encourage the planting of trees on marginal farmland. This action would have the additional advantage of reducing erosion and alleviating dust-bowl conditions in dry years. Planting trees along highway corridors would enhance their appearance. And all of these trees would be quietly removing carbon dioxide from the atmosphere while they added to the quality of our environment. We could say that these are beneficial "side effects" of a treatment to help reduce the threat of severe climate change.

At the same time, however, trees are being destroyed at an alarming rate in many parts of the world. Every year a territory the size of Tennessee is cleared of tropical forests. This activity accounts for about 20 percent of the current buildup of carbon dioxide. While it is relatively easy for us to undertake the planting of new trees in our own country, it is difficult to halt the destruction of trees in the undeveloped lands. Nevertheless, imaginative plans are being tried in a few test cases. Third-world nations, typically, are carrying heavy national debts that are a burden to their economy. A reduction of these debts offers a benefit that can be exchanged for a program of protecting natural areas and forest growth. A debt-for-nature exchange has recently been negotiated with Madagascar. The World Wildlife and Conservation Funds in cooperation with the United States Government are financing and helping to direct this project.

Madagascar has unique natural treasures that should be preserved for their own sake alone. This island separated from Africa 160 million years ago, carrying with it an assortment of plants and animals, which subsequently evolved in isolation. Today over 90 percent of Madagascar's fauna and 85 percent of its flora are found nowhere else in the world. These rare species are needed to maintain biological diversity, and a number of the plants secrete substances that have important medicinal applications.

But deforestation of the steep mountainsides is destroying this habitat for all living things. From the air, one can see that the ocean all around the island is stained a dark, bloody hue by the red soil washed down from the bare hills. Near the capital city of Antananarivo, the mountains are completely denuded of vegetation and are deeply indented with erosion channels. Here and there whole hillsides have slumped, leaving deep, red scars where the earth has fallen away.

In the city itself, the streets are awash with people: men pulling rickshaws, women with baskets on their heads, little girls carrying babies on their backs, boys guiding oxen pulling overloaded carts. Poverty is apparent everywhere—in the pitifully small shacks that serve as houses, in the overcrowded conditions, and the ragged clothes. Madagascar has too many people to be supported by this beautiful but dying land. They need space now occupied by the remaining forests to grow their food, and they cut the forest growth for fuel to cook their meals. The human population is increasing rapidly. As more people have to be fed, more forest land will be cleared. The debt-for-nature exchange is an attempt to break this vicious circle. An amount of 2.1 million dollars will be paid off the national debt in return for a nature-preservation program. Bolivia and the Philippine Islands have received similar aid in return for protecting their forests.

— • —

By planting at least as many trees as are felled every year and by halting the use of CFCs the accelerating rise of greenhouse gases in the atmosphere can be checked. In this way we can gain enough time to redirect our priorities and change our technology in ways that would reduce the burning of fossil fuels.

One of the best methods to achieve this goal is to make the use and production of energy more efficient. This option has great potential for reducing carbon dioxide emissions in a relatively short period of time, and recent history provides proof of its effectiveness.

In the wake of the OPEC oil embargo in 1973, conservation measures were enthusiastically adopted throughout the economy, and the savings were impressive. Before 1973, energy consumption in the United States had been increasing at an annual rate of 7 percent, a compound growth that would lead to a doubling of consumption in one decade. But from 1973 to 1984, energy consumption actually decreased slightly. This occurred during a period of strong economic growth; the gross national product (discounting inflation) rose more than 30 percent. These facts demonstrate the fallacy of a dogma that had been blindly accepted for many years—the belief that energy and economic growth go hand-in-hand in an industrial society. If one is reduced (so the theory went) the other must also decline. Now we know that neither economic growth nor standard of living are tied directly to energy consumption. There is another very large factor that affects the relationship—the efficiency of energy usage. It is not how much energy we demand but what it accomplishes for us that determines the quality of comfort and luxury of lifestyle.

Other industrial nations have done much better than the United States in their effective usage of energy. They have fewer natural resources and, therefore, have had more incentive to reduce the amount that they need to import. The

Germans, the Swiss, and the Japanese use about half as much as we do per capita, but their standard of living is at least as high.

During the energy crisis of the 1970s, considerable progress was made in the United States toward better levels of efficiency. Congressionally mandated fuel mileage standards produced cars that delivered more miles for each gallon of gas, and lower speed limits further increased efficiency, stretching the mileage achieved per unit of fuel. The general public cooperated because the price of gas had escalated and was in short supply. These conservation measures also produced a number of beneficial side effects— fewer deaths on the highways and reductions in air pollution.

At the same time, the cost of electricity was also rising steeply. In 1979, the average price per kilowatt hour (adjusted for inflation) was 250 percent higher than it had been before the oil embargo. As a result people did what executives in the energy business thought they would never do— they conserved. They turned out the lights when they left a room; they shopped for more efficient appliances; they installed insulation in their all-electric homes, and turned down their thermostats by a few degrees. This new policy of conservation took the power industry by surprise. They had assumed that electricity was not price sensitive, that people would continue to use more and more no matter what the price. Thus, another dogma was disproved. In 1981, the savings were so great that the residential sales of electricity were 105 billion kilowatt hours lower than projections based on historical trends. In the meantime the utilities had overbuilt, and the lower demand for power left them with unused capacity. So the price per kilowatt hour was raised again to cover the cost of facilities that were not running at full load.

Spurred by the new conservation philosophy, appliance

engineers were beginning to design equipment that did the same job with less power. Amana Refrigeration was a pioneer in this field. With a few simple changes, such as better insulation and more effective housing and sealing, they produced a standard-size refrigerator that used only 40 percent as much electricity as the previous model. For the owner of this appliance, the small, incremental original cost was paid back in reduced power bills in just a little over one year. Conservationists have estimated that if this saving could be multiplied by the 125 million refrigerators and freezers in the United States, the reduction would equal the output of twenty large power plants.

The case of the Amana refrigerator is just one example of thousands of possible engineering improvements that could lead to more efficient appliances. It is encouraging that in 1987 Congress passed an Appliance Energy Conservation Act requiring manufacturers of refrigerators, freezers, water heaters, washers and dryers, dishwashers, and air conditioners to meet specified targets for energy efficiency. This Act could save the public 28 billion dollars in electricity by the end of the century and eliminate the need for twenty-two new power plants. Since each coal-burning 1,000 megawatt power plant emits approximately 6 million tons of carbon dioxide a year, this one act alone could reduce emissions by 100 million tons annually.

Many conservation measures have been initiated but have not been universally adopted. Recycling, for example, has great potential for saving energy. The processing of aluminum is very energy-intensive, and recycling cuts this energy cost in half. The recycling of newspapers reduces the number of trees that must be felled to produce the paper, trees that otherwise could be helping to control the greenhouse effect. Some plastic containers can be recycled, thereby reducing the expense of waste disposal and preventing the accumulation of unsightly trash that will not deteriorate for

100 years or more. Other almost untapped possibilities include more efficient lighting, better architectural design, and many options for industrial processes.

— • —

Looking back, we realize that we should be thankful for the energy crisis of the 1970s, which introduced these methods of conservation to us. It started a swing toward energy efficiency that has reduced fossil-fuel consumption, now growing at a much slower rate than before the embargo. Thanks to the discovery of conservation as a viable economic policy, there is potential for reducing this rate further or even eliminating its growth entirely by greater efficiency.

Amory Lovins presents a very optimistic scenario based on this principle in his book *Least-Cost Energy: Solving the CO_2 Problem*. He states that "economically efficient energy investments will at least hold constant, and probably reduce the world rate of burning of fossil fuel, starting immediately, even assuming rapid worldwide economic growth and complete industrialization of the developing countries."

Unfortunately, the conservation movement lost some of its momentum in the mid-1980s. The effort to become less dependent on imported oil had been so successful both here and abroad that the OPEC nations found the demand for their oil greatly reduced—from 31 million barrels a day in 1981 to 17 million in 1985. As a result, they marked down the price of oil, and the conservation trend began to reverse itself. In the United States, cheap and abundant gasoline encouraged the buying of larger, less-efficient cars. And these were driven faster on the highways. Yielding to public pressure, Congress increased speed limits on the highways and relaxed the requirement for improved mileage averages in the automotive fleets. Twenty-seven and one-half miles per gallon is now the mandated level (Detroit said it would be very difficult to achieve higher standards.) But many fuel-efficient cars have been built in Europe. In Sweden, Volvo has designed a car that delivers sixty miles per gallon. Of course, in Europe gasoline is much more expensive—$2.50 to $4 a gallon. A large part of this is tax, so the price is not as sensitive to changes in OPEC prices.

The lesson is obvious: The public cooperates in conservation measures if the price they pay for energy is a fair measure of its cost. A high price indicates a valuable commodity and people use it with care. They compensate for larger unit costs by demanding greater efficiency in their appliances and cars and by taking personal measures to use it economically. If the price of energy included the expense of measures to control environmental damage caused by its usage, then the public would have an honest measure of its *real cost* and would work actively toward achieving greater savings.

One of the reasons for the wastefulness that has become typical of energy usage in the United States is the deceptively small price tag for units of energy. Not only has the cost of environmental and health consequences been passed through, but substantial subsidies have made energy look cheaper than it really is. Lovins estimated in 1981 that current tax and price subsidies "total over 100 billion dollars a year, enough to reduce the average energy price by over a third and nuclear energy price by over a half." The development of nuclear energy has been heavily subsidized and promoted by the government. These inequalities disturb the operation of a free-market economy. They are giving the consumer false signals, and he is denied the option of adjusting his energy usage based on real cost.

If the market were not artificially manipulated in this way, alternative energy sources might be brought to practical realization much more rapidly. Solar energy, for example, has not been able to compete with electricity on a unit-cost basis, but if electric rates included the extra expense of avoiding unnecessary damage to the environment, then these systems would soon become economically competitive and their development would be commercially attractive.

Solar energy is almost pollution free; it produces no greenhouse gases, no sulfur or nitrogen oxides, no hydrocarbons. The energy is free, although unevenly distributed and intermittent. The solar technology known as *photovoltaics* appears to have the best potential for displacing significant amounts of fossil-fuel generation.

Solar cells, which turn sunlight directly into electricity, were invented at Bell Laboratories in the 1950s. These are made of semiconductor material whose atoms absorb light, freeing electrons and creating "holes" that carry the current. Each cell has two or more layers of semiconductor materials that have slightly different characteristics and, when they are connected, a voltage is created, driving the electrons through the circuit.

The first solar cells were perfected for use in the space program. In this application, reliability was the most important consideration; so these early solar cells were custommade and were extremely expensive. However, they did prove to be dependable and, once the original cost had been paid, they ran year after year on free energy.

The development of less expensive solar cells was undertaken during the 1970s in the hope that this technology could displace a significant proportion of oil-fired electric generation. Some workers dreamed of making solar cells so cheap that rooftop installations could provide power directly for homes and small businesses. Government funding helped to finance this research, although the support for solar power never approached that of nuclear power. In spite of remarkable advances in efficiency and cost reduction, the goals have not yet been achieved, and government funding has been cut back to the point that it is a negligible factor.

However, it is a wonderful testimonial to the great potential of this technology that, in spite of policies favoring other sources of energy, the price of photovoltaic systems today is approaching a level that is competitive with some types of conventionally generated electricity.

Silicon crystals were the first semiconductors used in solar cells, and they still dominate the market. Production methods taking advantage of semiautomation have steadily reduced the cost of these crystals, and new understanding of semiconductors has led to higher levels of efficiency. Research cells made of silicon crystals have reached an efficiency of 22.8 percent in direct sunlight. Methods of concentrating the sunlight that falls on the solar cells improves their efficiency although it adds to the expense of construction.

Another approach to the challenge of reducing the cost of a photovoltaic system is to sacrifice some efficiency in the semiconductors in return for lower production costs. Many different types of semiconductor material are under investigation: ribbons of polycrystalline silicon, amorphous silicon, thin films of other materials such as copper and cadmium compounds, gallium arsenide and its alloys.

The research effort has made impressive progress. Over the last decade, the costs of solar-cell-generated power has been reduced fortyfold, and there is potential for considerably greater reduction.

Even under the present price structure, photovoltaic technology should be competitive for generating central-station peaking power by the late 1990s. On hot summer days when airconditioners are running at full power, additional electricity is needed to meet the demand. This peaking power is somewhat more expensive than base load to provide, and it represents about 17 percent of installed capacity. Fortunately, solar energy is also at a maximum during these hours.

Workers in the field of photovoltaics believe that this source of power will be a viable option to meet many of the electrical requirements in the United States within the next four or five decades. If global warming should become a serious problem, the facilities needed to provide this pollution-free power could be built quite rapidly. A large photovoltaic plant can be brought on line in just one or two years.

— • —

The uneven distribution of the Sun's energy on Earth has been one of the stumbling blocks on the road to fuller reliance on solar power. Reliable storage and transmission systems are necessary to overcome this problem. One attractive possibility is a technique that has long been used to obtain pure hydrogen. By passing electricity through water containing a small amount of electrolyte, water is split into its primary elements: two parts hydrogen and one part oxygen. When these elements are recombined, energy is given off, energy that can be used to generate electricity. Thus a closed cycle is obtained, which provides a method of storing power.

Alternatively, the hydrogen produced by off-peak electricity in large solar plants could be piped underground to cities in areas that receive less abundant sunlight. There, it could be used to replace petroleum fuels. Hydrogen burns very cleanly; the only by-products are a small amount of nitrogen oxides and water vapor made when hydrogen combines with the oxygen in the air.

Although hydrogen is highly flammable, engineers believe that with proper design it could be used safely to fuel transportation systems and home heating. Electricity could be generated in the home by piping hydrogen into electrochemical fuel cells that convert it directly into electric current.

At the present time, these possibilities have not been commercially developed because the price differential is too great. But if a real-cost policy were pursued this differential might disappear. The cost of pollution from gasoline, for example, has been estimated at about one dollar per gallon, and if the price were raised to include this cost, hydrogen fuel would be approximately competitive.

— • —

One alternative to fossil fuel as an energy source is, of course, nuclear power. Back a few decades ago when scientists began to talk about the potential danger of greenhouse warming, the advocates of nuclear power enthusiastically pursued this issue, hoping that it would help to counteract the growing public resistance to nuclear technology. Fears of exposure to radioactive elements in case of a malfunction or melt-down and the persistent unsolved problem of radioactive-waste disposal were causing the American public to resist the building of new nuclear plants. However, it could be maintained that these concerns were based on hypothetical disasters that might never materialize. If the fear of a serious climate change caused by continued burning of fossil fuels appeared to be based on more realistic assumptions, it might overcome the objections to nuclear power. Perhaps without Three Mile Island and without Chernobyl this argument would have carried the day. But, after these incidents, the public recognized that the danger of nuclear energy was not hypothetical; it was caused by a factor that could never be corrected—human error.

Furthermore, and this was probably the overriding consideration, nuclear power did not turn out to be the bargain that had been expected. The plants proved to be very expensive and their construction took at least ten years.

Even if a thousand new nuclear plants were ordered today,

the impact on the buildup of greenhouse gases would be small and too late as compared to the immediate effect offered by conservation and by reforestation.

— • —

It is possible, however, that the energy embodied in the atom could provide vast resources of power without the hazards associated with the present nuclear technology. The plants operating today depend on fission, the splitting of large atoms like uranium. In this reaction, energy is given off as well as small atomic particles that go on to cause the splitting of other atoms; so the process becomes a chain reaction. Many of the products are radioactive, some lasting as long as several hundred thousand years.

However, there is another way energy can be released from the atom. This is known as *fusion*, the joining of two light atoms to make a larger atom. When hydrogen atoms come together to make helium, for example, vast amounts of energy are released and this fusion can also become a self-sustaining chain reaction. The process has been understood for many years and is, in fact, the basis of the energy produced in the hydrogen bomb. In this case, a small fission bomb is used as a trigger to produce the enormous amount of energy required to force the nuclei of the hydrogen atoms close enough together to start the fusion process. The reaction takes place with an explosive force that cannot be controlled. But, of course, in a bomb this is just the effect that is desired.

If this force could be harnessed in a way that could be regulated so the energy is given off slowly enough to be safely under control, then an almost unlimited source of power would be available. The isotopes of hydrogen that would be used as fuel are widely available and the supply is nearly inexhaustible. The products of this reaction are not radioactive with the exception of tritium—one of the hydrogen isotopes—and tritium may be usable as a fusion fuel. Therefore, it would not be left over at the end of the process, and no radioactive wastes would result.

Atomic physicists have been working to meet this challenge for several decades, but so far they have been unable to achieve a chain reaction that produces more energy than is initially fed into the system to force the atoms to come together. The hydrogen nuclei are positively charged and, therefore, they repulse each other. With enormous pressures and magnetic fields this repulsion can be overcome.

Another, totally different possibility was announced in the spring of 1989. Stanley Pons working at the University of Utah and Martin Fleischmann at the University of Southhampton in England had cooperated on a very simple little experiment. A typical laboratory beaker was filled with a special kind of water made with deuterium, one of the isotopes of hydrogen. This "heavy" water was then subjected to electrolysis, using an electric current flowing between two electrodes—one of platinum and one of palladium, a metal that is sometimes used in jewelry as a substitute for platinum. A small amount of lithium hydroxide was dissolved in the water to act as an electrolyte. To the surprise and delight of the researchers, the experiment produced more energy in the form of heat than had been fed into the cell as electric current, and more than could be accounted for by any known chemical reaction.

The announcement of these results was greeted with great skepticism by atomic physicists who had been working on the other methods of producing fusion. A rush to reproduce the experiment led to frustration and further skepticism. Sometimes it could be repeated—but often it failed. The scientists who made the original discovery frankly admitted that they did not understand exactly what was happening and did not know the precise conditions that must be met in order for this reaction to be sustained. As of this writing, many unanswered questions still remain. Is this really a form of fusion? If not, what is the source of the extra energy?

By an interesting coincidence, at the same time as the original University of Utah experiment, laboratory work using similar equipment was producing positive results at Brigham Young University. Steven E. Jones had detected what he believed to be fusion at room temperature, but at a much slower rate than Pons and Fleischmann had claimed.

If indeed fusion is occurring—even at a very slow rate—without the use of any extreme measures such as the application of enormous pressures, temperatures, or magnetic fields, then this is a very important discovery. It would affect all the theories of the formation of matter and perhaps even the creation of the universe itself.

In a very general way, we can understand how a process like this could occur when we recognize that it is probably a catalytic phenomenon. Palladium is known to be an effective catalyst in other reactions, and we know that catalysts act in a way that seems to be pure magic, allowing changes to take place at "normal" temperatures and pressures that would otherwise occur only under extreme conditions. For example, the fixing of nitrogen requires enormous amounts of energy, such as those delivered by a stroke of lightning, but the tiny bacteria living on the roots of legumes produce the catalyst that fixes nitrogen in darkness at ambient temperatures.

We have encountered phenomena of this type over and over again throughout these pages. So we should not be totally surprised to find that nature makes use of a catalytic method to build matter, fusing lighter atoms to make more complex ones.

Whether or not "cold fusion" could be used any time soon to produce more energy for mankind is another question. Years of further research may be required before this question can be answered. We cannot wait for such attractive options, tempting as the prospect may be. We must use the technologies at hand to meet the problems we ourselves have created before they become too massive to control. The solution will require all nations working together, because these are worldwide problems that cannot be solved by any one nation alone. This need to cooperate with other countries poses a great challenge to man's adaptability, but like all challenges it offers an important opportunity. In the face of a common enemy, we may be inspired to put aside our regional disputes and our jockeying for power—to pause, at least for a moment, in the stockpiling of vast arsenals for mutual destruction, and to unite in a joint effort to restore the fine, blue layer of atmosphere that has protected, nurtured, and delighted mankind for at least three million years.

This effort cannot be left solely to the leaders of nations—to the politicians and generals and presidents. It must spring from the common commitment of the vast majority of people, people making individual contributions day by day. The recent events in Eastern Europe have demonstrated the enormous power that people, working together, can wield. Without the use of military force, the Iron Curtain has been been torn asunder and the Berlin Wall has been breached. The people are in the process of creating a new world, and the leaders have followed. The movement to preserve our environment need not be so dramatic but it must be sustained throughout every lifetime. Little economies must be practiced in the use of energy, in the reduction of waste, and in the recycling of materials. These are small housekeeping chores like those we do every day to maintain, restore, and embellish the small piece of Earth that we call our homes. If we think of the whole Earth as our home in space we can work together effectively to shield it from the seeds of change.

The stakes are high because, if we should fail, our grandchildren will have to retreat more and more to an artificial indoor environment. And outdoors they will find a very different world than the one we have enjoyed. There will be fewer days when the Sun shines in a cloudless sky and the Earth shines back, returning the light in a spectrum of radiant color—the vibrant green of young wheat fields in the spring, the glowing reds of bougainvillea and hollyhocks, the golden yellow of aspen leaves that gladdens the heart like laughter. In that world of tomorrow, the sky will be darkened by an all-pervasive smog changing the blues to greys, draining the color from the aqua seas, and obscuring the purple mountain ranges capped with snow. Will our grandchildren ever know the sudden glory of a dawn in an unpolluted sky or witness the slow fading of the violet light that floods a clear twilight heaven? Will they explore a mountain meadow knee deep in columbine or walk in the tranquil aisles of a virgin forest where the trees soar tall and proud to meet the Sun?

"There was a child went forth," Walt Whitman said, and everything the child saw became part of him: "The early lilacs became part of this child/ and grass, and the white and red morning glories, and white and red clover, and the song of the phoebe-bird." As these sights and sounds are replaced by cold stone and hard steel and plumes of acrid smoke, then these will become part of the child. A fabric woven of such coarse threads will make a harsher man.

The mobile membrane of air that hugs the curving surface of our planet is a masterpiece of nature's art, embroidered with high-flying cirrus and rainbows, animated with bird song and the long, liquid arc of the swallow's flight. If we sully this delicate halo of air, we change the nature of every living thing that shares this beautiful planet with us. And we are just as surely changing the nature of man.

References

INTRODUCTION

Page

xii *Lake Michigan levels:* Data supplied by Illinois State Water Survey.

xiii *New information on lake levels:* Curtis E. Larsen, "A stratigraphic study of beach features on the southwestern shore of Lake Michigan: new evidence of Holocene lake level fluctuations": *ISGS Environmental Geology Notes 114,* 1986.

xv *Eddington quotation:* Cited by Tobias Dantzig in *Number: The Language of Science.* New York: The Macmillan Company, 1944.

CHAPTER 1

5 *Astronomical distances:* Personal communication with Dr. Eric Carlson, The Adler Planetarium, Chicago, Illinois, September 21, 1989. These distances are approximations only and are constantly being altered as new information is discovered.

7 *Montgolfier balloon:* C. V. Glines, ed., *Lighter-than-Air Flight.* New York: Franklin Watts, 1965.

8 *Andree's Flight:* C. V. Glines, ed. *Lighter-than-Air Flight.* New York: Franklin Watts, 1965, pp. 78–81.

12 *Glaisher quotation:* From J. Glaisher, C. Flammarion, W. Fonvielle, G. de Tissandier, "Coxwell and Glaisher's Dangerous Ascent," *Voyages Aeriens.* Paris, 1870.

13 *Glaisher and Coxwell flight:* For the details of this flight I have followed the story as told by Stehling and Beller in *Skyhooks.* Garden City, N.Y.: Doubleday, 1962.

14 *First Simons quotation:* David G. Simons with Don A. Schanche, *Man High.* Garden City, N.Y.: Doubleday, 1960, p. 135.

14 *Second Simons quotation:* David G. Simons, "Man High," *Life,* (October 21, 1957), pp. 19–27.

CHAPTER 2

23 *Lewis Thomas quotation:* Lewis Thomas, *The Lives of a Cell.* New York: Viking Press, 1974, p. 43.

28 *Japanese balloon offensive:* Lincoln LaPaz with Albert Rosenfeld, "Japan's Balloon Invasion of America," *Collier's,* January 17, 1953, pp. 9–11; and W. H. Wilbur, "Those Japanese Balloons," in C. V. Glines, ed., *Lighter-than-Air Flight.* New York: Franklin Watts, 1965.

37 *Gliding contest in Germany:* Leo Loebsack, *Our Atmosphere.* Translated by E. L. and D. Rewald. New York: New American Library, 1961, p. 122.

38 *Measurements of high cloud cover:* Reported in "Is Man Changing the Earth's Climate?" *Chemical and Engineering News,* August 16, 1971, pp. 40–41. And John Horgan, "Pinning Down the Clouds," *Scientific American* (May 1989), p. 24.

39 *University of Chicago study:* Reported by Tim Obermiller, "A Delicate Balance," *University of Chicago Magazine* (Spring 1989), pp. 14–20.

40 *Volcanic eruptions and colder weather:* Correlation as discussed by Schneider and Mass in "Volcanic Dust, Sunspots, and Temperature Trends," *Science,* Vol. 190 (November 21, 1975), pp. 741–742. Evidence of poor growing seasons and freezes following volcanic eruptions was found by scientists at the University of Arizona's tree-ring laboratory, as reported by *Time* (March 24, 1975), p. 64.

40 *Volcanic activity and the ice ages:* Evidence of correlation from deep-sea cores, cited by B. Baldwin, James B. Pollack *et al,* "Stratospheric Aerosols and Climatic Change," *Nature,* Vol. 263, 5578 (1977), pp. 551–555.

42 *Shakespeare quoted: Hamlet,* Act II, Scene 3.

CHAPTER 3

48 *Sensitivity of organisms to oxygen:* Preston Cloud and Aharon Gibor, "The Oxygen Cycle," *Scientific American*, Vol. 223, No. 3 (September 1970), pp. 110–123.

49 *Timing of oxygen levels concentrations:* Personal communication from J. John Sepkowski, Jr., Professor of Geophysical Sciences, University of Chicago, September 13, 1989.

56 *Nitrogenase:* C. C. Delwiche, "The Nitrogen Cycle," *Scientific American*, Vol. 223, No. 3 (September 1970), p. 142.

60 *Schaefer and Langmuir experiments:* Described by J. Gordon Cook in *Our Astonishing Atmosphere*. New York: Dial Press, 1957, pp. 128–31.

61 *Measurements of sunshine on earth's surface:* Reported in *The Wall Street Journal*, Feb. 20, 1975.

61 *La Porte weather changes:* Cited in *McGraw-Hill Encyclopedia of Science and Technology*, Vol. 3. New York: McGraw-Hill Book Co., 1971, p. 185.

62 *St. Louis experiment:* Roscoe R. Braham, Jr., "Overview of Urban Climate," *Proceedings of Conference on Urban Physical Environment*, Syracuse, N.Y., August 25–29, 1975, and in "Cloud Physics of Urban Weather Modification—A Preliminary Report," *Bulletin of the American Meteorological Society*, Vol. 55 (February 1974), pp. 100–106.

65 *El Niño:* Kevin E. Trenberth, Grant W. Branstator, and Phillip A. Arkin, "Origins of the 1988 North American Drought," *Science*, Vol. 242 (December 23, 1988), pp. 1640–1645.

66 *Deep-sea vents:* Walter Sullivan, "Deep-Sea Life is Found Flourishing on Sulfur From Ocean's Volcanoes," *The New York Times*, April 25, 1982, p. 33.

67 *Quantities of heat and water emerging from the vents:* Mitch Waldrop, "Ocean's Hot Springs Stir Scientific Excitement," *Chemical and Engineering News*, March 10, 1980, pp. 30–33.

67 *Muir quotation:* John Muir, *Gentle Wilderness*. San Francisco: Sierra Club Books, p. 146.

CHAPTER 4

71 *Lewis Thomas quote:* From Lewis Thomas, *The Lives of a Cell.* New York: Viking Press, p. 145.

72 *Thickness of ozone layer if brought to sea levels:* J. A. Ratcliffe, *Sun, Earth, and Radio.* New York: McGraw-Hill, 1970, p. 25.

73 *El Fuego dust cloud:* M. P. McCormick and W. H. Fuller, Jr., "Lidar Measurements of Two Intense Stratospheric Dust Layers," *Applied Optics,* Vol. 14, No. 1 (January 1975), p. 4.

73 *Circulation patterns in the stratosphere: Environmental Impact of Stratospheric Flight,* report of the Climatic Impact Committee of the National Research Council, the National Academy of Sciences and the National Academy of Engineering. Washington, D.C.: National Academy of Sciences, 1975, p. 6.

73 *Variations in density of ozone layer:* "Fluorocarbons and the Environment," report of the Federal Task Force on Inadvertent Modification of the Stratosphere, Council on Environmental Quality and Federal Council for Science and Technology, June 1975, p. 69.

75 *Experiment on Noctilucent clouds:* Robert K. Soberman, "Noctilucent Clouds," *Scientific American,* Vol. 208, No. 6 (June 1963), pp. 51–59.

79 *Description of Aurora:* Syun-Ichi Akasofu, "The Dynamic Aurora," *Scientific American* (May 1989), pp. 90–97. Also Thomas Potemra, "The Aurora!" *Johns Hopkins Magazine* (October 1980), pp. 8–13.

80 *Aurora and sunspot cycles:* E. N. Parker, "The Sun," *Scientific American,* Vol. 233, No. 3 (September 1975), pp. 43–50. Also John A. Eddy, "The Maunder Minimun," *Science,* Vol. 192 (June 18, 1976), pp. 1189–1202.

81 *Ladurie's study of Alpine glaciers:* Reported by Henry Lansford in "Climate Outlook: Variable and Possibly Cooler," *Smithsonian* (November 1975), p. 144.

81 *Geologists find evidence of glacial growth:* Jean M. Grove, *The Little Ice Age.* New York: Methuen Inc., 1988.

81 *Midwest temperatures in 1830s:* "Report of the Ad Hoc Panel on the Present Interglacial," Federal Council for Science and

Technology, National Science Foundation (August 1974), p. 20.

82 *Twelfth Century solar activity:* Personal communication from Eugene S. Parker, Professor of Astronomy and Physics at the University of Chicago, Sept. 5, 1989.

83 *Condition of the bodies of the Vikings:* Stephen H. Schneider and Randi Londer, *The Coevolution of Climate and Life.* San Francisco: Sierra Club Books, 1984, p. 112.

83 *Viking settlements in Greenland:* Reid A. Bryson and Thomas J. Murray, *Climates of Hunger,* University of Wisconsin Press, 1977, pp. 67–71. Also Fradley Garner and Jens Rosing, "Lost Norse Mystery," *Oceans,* Vol. 10 (March 1977), pp. 4–9.

83 *Conditions in Greenland today:* Personal communication from Eugene Parker, Professor of Astronomy and Physics, University of Chicago, Sept. 5, 1989.

84 *Parker quoted:* Personal communication, July 19, 1989.

89 *Milankovitch theory:* J. D. Hays, John Imbrie, and N. J. Shackleton, "Variations in the Earth's Orbit: Pacemaker of the Ice Ages," *Science,* Vol. 194 (December 10, 1976), pp. 1083–1087.

90 *Speed of changes between glacial and interglacial:* Reid A. Bryson and Thomas J. Murray, *Climates of Hunger,* University of Wisconsin Press, 1977, pp. 126–129.

CHAPTER 5

96 *Concentration of carbon dioxide: Physical Monitoring for Climate Change,* No. 16, Summary Report 1987, U. S. Department of Commerce, National Oceanic and Atmospheric Administration, Environmental Research Laboratories, p. 31.

96 *Long wave radiation:* William W. Kellogg, "Is Mankind Warming the Earth?" *Bulletin of the Atomic Scientists* (February 1978), pp. 13–14.

97 *Weather in Europe and America:* Stephen Schneider, *The Genesis Strategy.* New York: Plenum Press, 1976, pp. 70–78.

98 *Northern migration of fish and birds:* Rachel Carson, *The Sea Around Us.* Oxford University Press, 1951, pp. 183–186.

98 *Temperature changes in Northern Hemisphere:* "Report of the Ad

Hoc Panel on the Present Interglacial," Federal Council for Science and Technology, Science and Technology Policy Office, National Science Foundation, August, 1974, p. 20.

98 *Hubert Lamb research:* Cited by Henry Lansford, "Climate Outlook: Variable and Possibly Cooler," *Smithsonian* (November 1975), p. 142.

98 *Temperature changes in North Atlantic:* Roger G. Barry, John T. Andrews, and Mary A. Mahaffy, "Continental Ice Sheets: Conditions for Growth," *Science,* Vol. 190 (December 5, 1975), p. 980.

99 *Volcanic eruptions and climate:* Richard A. Kerr, "Volcanoes Can Muddle the Greenhouse," *Science,* Vol. 245 (July 14, 1989), pp. 127–128.

99 *Decline in transparency:* R. A. Bryson and G. J. Dittberner, "A non-equilibrium model of hemispheric mean surface temperature." *Journal of the Atmospheric Sciences,* Vol. 33 (1976), pp. 2094–2106.

99 *Man-made dust comparable to moderate volcanic activity:* Reid A. Bryson, "A Perspective on Climatic Change," *Science,* Vol. 184 (May 17, 1974), pp. 753–760.

100 *United Nations study:* Cited by the *The New York Times,* March 7, 1976.

100 *Reduction of temperature by dust veil:* Reid A. Bryson, "A Perspective on Climatic Change," *Science,* Vol. 184 (May 17, 1974), p. 758.

101 *Media descriptions of greenhouse effect:* This material is a composite of statements that have appeared recently in the press.

102 *Mark Twain quote:* Mark Twain, *Life on the Mississippi.* New York and London: Harper & Brothers, 1917, p. 136.

102 *Schneider quote:* Stephen H. Schneider and Randi Londer, *The Coevolution of Climate and Life.* San Francisco: Sierra Club Books, 1984, p. 232.

104 *Possible increase in global temperature:* Stephen H. Schneider, "The Greenhouse Effect: Science and Policy," *Science,* Vol. 243 (February 10, 1989), p. 774.

105 *The Altithermal:* William W. Kellogg, "Is Mankind Warming the Earth," *Bulletin of the Atomic Scientists* (February 1978), pp. 11–19.

105 *Analog of the Altithermal:* Reid A. Bryson, "On Climatic Analogs in Paleoclimatic Reconstruction," *Quaternary Research,* Vol. 23 (1985), pp. 275–286.

106 *Global warming in the last 100 years:* Stephen H. Schneider, "The Greenhouse Effect: Science and Policy," *Science,* Vol. 243, No. 4892 (February 10, 1989), p. 772.

106 *Temperatures in the forty-eight contiguous states:* Philip Abelson, "Climate and Water," *Science,* Vol. 243, No. 4890 (January 27, 1989), p. 461.

108 *Information from ice cores:* Jonathan Weiner, "Glacier Bubbles are Telling Us What Was in Ice Age Air," *Smithsonian* (May 1989), pp. 84–88.

109 *Volcanic activity from 1880 to 1912:* Owen B. Toon and James B. Pollack, "Volcanoes and the Climate," *Natural History,* Vol. 86, No. 2 (January 1977), p. 26. Also Richard A. Kerr, "Volcanoes Can Muddle the Greenhouse," *Science,* Vol. 245 (July 14, 1989), pp. 127–128.

109 *Gaia hypothesis:* James E. Lovelock, *Gaia: A New Look at Life on Earth.* Oxford University Press, 1979.

113 *Diversity of Amazon rain forest:* Stephen Schwartzman, "Protecting the Amazon Rain Forest, A Global Resource," *EDF Letter,* A report to Members of the Environmental Defense Fund, Vol. XX, No. 1 (February 1989), p. 7.

113 *Satellite photographs of Amazon forest:* Stephan Schwartzman, "Protecting the Amazon Rainforest, A Global Resource," p. 7.

114 *Extractive reserves: EDF Letter,* p. 7.

114 *Assassination of Francisco Mendes:* Reported in *EDF Letter,* pp. 1, 7; also Mason Willrich, "Murder in Acre, *The Amicus Journal* (Spring 1989), p. 10.

114 *Sarney's announcement of reforms:* Reported in *EDF Letter,* p. 1.

114 *Addition of carbondioxide by deforestation:* Richard A. Houghton and George Woodwell, "Global Climate Change," *Scientific American,* Vol. 260, No. 4 (April 1989), p. 41.

115 *Cows' production of methane:* George Nobbe, "Belchless Cows," *Omni* (May 1989), p. 36.

116 *Particulate concentration over land and ocean:* Personal communication by V. Ramanathan, Professor of Atmospheric Sciences, University of Chicago.

117 *Cooling effect of cloud cover:* V. Ramanathan *et al*, "Cloud-Radiative Forcing and Climate: Results from the Earth Radiation Budget Experiment," *Science*, Vol. 243 (January 6, 1989), p. 57.

117 *Current models assume positive cloud feed-back:* Richard A. Kerr, "How to Fix the Clouds in Greenhouse Models," *Science*, Vol. 243 (January 6, 1989), p. 29.

118 *Spreading dust in the stratosphere:* Cited by Stephen H. Schneider, "The Greenhouse Effect: Science and Policy," *Science*, Vol. 243 (February 10, 1989), p. 775.

118 *Cover the oceans with Styrofoam chips:* Reported by William J. Broad in "Scientists Dream up Bold Remedies for Ailing Atmosphere," *The New York Times*, August 16, 1989, p. 19.

CHAPTER 6

121 *Forests in Germany and Czechoslavakia:* Statistics on forest death compiled by West Germany's Interior Ministry in 1982, cited by Lon Luoma, "Dead Forests and Acid Bananas," *Audubon* (September 1983), p. 38.

122 *Lakes in Norway and Sweden:* Robert H. Boyle and R. Alexander Boyle, "Acid Rain," *The Amicus Journal* (Winter 1983), p. 24.

123 *Chemistry of acid rain:* Volker A. Mohnen, "The Challenge of Acid Rain," *Scientific American*, Vol. 259, No. 2 (August 1988), pp. 30–38. In addition to the principal reactions described here there are other reactions involved in the formation of acids from nitrogen and sulfur oxides.

124 *Concentration of sulfur and nitrogen compounds found in ice cores:* Jonathon Weiner, "Glacier Bubbles Are Telling Us What Was in the Ice Age Air," *Smithsonian* (May 1989), p. 84.

124 *History of acid rain as an environmental concern:* Stephen E. Schwartz, "Acid Deposition: Unraveling a Regional Phenomenon," *Science*, Vol. 243, No. 4892 (February 10, 1989), p. 754.

124 *Dying lakes in New England and Canada:* Robert H. Boyle and R. Alexander Boyle, "Acid Rain," *The Amicus Journal* (Winter 1983), p. 24.

125 *Experimental acidification of a lake in Canada:* Robert H. Boyle and R. Alexander Boyle, "Acid Rain," *The Amicus Journal* (Winter 1983), p. 27.

126 *Lake George:* Tom Kelly, "How Many More Lakes Have to Die," *Canada Today/D'Aujourd'Hui*, Vol. 12, No. 2 (February 1989), Published by the Canadian Government, p. 6.

127 *Die-off in acidified lakes:* Robert H. Boyle and R. Alexander Boyle, "Acid Rain," *The Amicus Journal* (Winter 1983), pp. 29–30.

128 *Brookhaven study:* Robert H. Boyle and R. Alexander Boyle, "Acid Rain," The Amicus Journal (Winter 1983), p. 30.

128 *Vogelmann's impressions:* Quoted by Jon R. Luoma, "Dead Forests and Acid Bananas," *Audubon,* September 1983, p. 38.

129 *Statement of AEP engineer:* Quoted by Jon R. Luoma, "Dead Forests and Acid Bananas," *Audubon,* September 1983, p. 40.

130 *Possibilities of damage to trees from acid rain:* Volker A. Mohnen, "The Challenge of Acid Rain," pp. 34–35. Also E. D. Schulze, "Air Pollution and Forest Decline in a Spruce (*picea abies*) Forest," *Science,* Vol. 244 (May 19, 1989), pp. 776–783.

131 *Statements by industry spokesmen:* American Electric Power Company advertisement quoted by Robert H. Boyle and R. Alexander Boyle, "Acid Rain," *The Amicus Journal* (Winter 1983), p. 26.

132 *Haziness caused by sulfate droplets:* Volker A. Mohnen, "The Challenge of Acid Rain," *Scentific American,* Vol. 254, No. 2 (August 1988), p. 31.

132 *Satellite photos:* Stephen E. Schwartz, "Acid Deposition: Unraveling a Regional Phenomenon," *Science,* Vol. 243, No. 4892 (February 10, 1989), p. 755.

132 *Sulfur dioxide emissions in Ohio and the Northeast:* Stephen E. Schwartz, "Acid Deposition: Unraveling a Regional Phenomenon," *Science,* Vol. 243, No. 4892 (February 10, 1989), p. 754.

132 *Newspaper articles about Ohio's role in acid clean-up: Columbus Dispatch,* June 8, June 9, and July 2, 1989.

133 *National Academy of Sciences study:* Reported in *The New York Times,* June 29, 1983, p. 9.

133 *Oppenheimer quote:* Cited by *EDF Letter,* a report to members of the Environmental Defense Fund, November, 1985, pp. 1–3.

133 *Study made in western states:* Michael Oppenheimer, Charles B. Epstein, and Robert E. Yuhnke, "Acid Deposition, Smelter Emissions and the Linearity Issue in the Western States," *Science,* (August 30, 1985), pp. 859–862.

134 *Sudbury:* Robert H. Boyle and R. Alexander Boyle, "Acid Rain," *The Amicus Journal* (Winter 1983), p. 32.

135 *Auto makers fear higher costs:* Doron P. Levin, "Auto Makers' Plea on Pollution," *The New York Times,* July 21, 1989, p. 25.

135 *Acid-rain in Africa:* Marlise Simons, "High Ozone and Acid-Rain Levels Found Over African Rain forests," *The New York Times,* June 19, 1989, p. 1.

135 *Carbon monoxide from savannah burning:* Reginald Newell *et al,* *Scientific American,* Vol. 261, No. 4 (October 1989), pp. 82–90.

138 *Nitrogen oxide removal:* New system announced in *Logos,* Argonne National Laboratories, Vol. 7, No. 1, pp. 9–13.

139 *TVA rate increases:* Reported in *NRDC Newsline* (Summer 1983), p. 1.

139 *Schwartz quotation:* "Stephen E. Schwartz: Unraveling a Regional Phenomenon," *Science,* Vol. 243, No. 4892 (February 10, 1989), p. 761. Emphasis is mine.

141 *Reports of astronauts:* Cited by Ernst Messerschmid in *The Home Planet,* Kevin W. Kelley, ed., for the Association of Space Explorers. Reading, Mass.: Addison-Wesley Publishing Company, 1988, p. 71.

CHAPTER 7

146 *Byrd quotation:* Richard E. Byrd, *Alone.* Covelo, Calif.: Island Press, 1984.

148 *Low readings taken at British station and by Nimbus 7:* Ellen Ruppel Shell, "Solo Flights into the Ozone Hole," *Smithsonian* (February 1988), pp. 142–155.

148 *Hole affecting South America and Australia:* Personal communication with John E. Frederick, Professor of Atmospheric Sciences, University of Chicago, July 19, 1989. Also as reported in *Scientific American,* Vol. 261, No. 4 (October 1989), p. 26.

148 *Amount of CFCs expelled into the atmosphere:* "Fluorocarbons

and the Environment," report of Federal Task Force on Inadvertent Modification of the Stratosphere, Council on Environmental Quality and Federal Council for Science and Technology, June, 1975, p. 69.

149 *Diffusion and residence time for chlorine atoms:* Figures cited in "Chlorofluorocarbons Threaten Ozone Layer," *Chemical and Engineering News,* September 23, 1974, pp. 28–29.

150 *Estimates of time delay in restoring ozone layer:* Steven C. Wofsy, Michael B. McElroy, and Nien Dak Sze, "Freon Consumption: Implications for Atmospheric Ozone," *Science,* Vol. 187 (February 14, 1975). Also *Halocarbons: Effects on Stratospheric Ozone,* report of Panel on Atmospheric Chemistry, Committee on Impacts of Stratospheric Change, National Academy of Sciences, Washington, D.C., September, 1976), pp. 1–23.

150 *Effect of nuclear war on ozone layer:* Reported by the National Academy of Sciences, "Long-Term Worldwide Effects of Multiple Nuclear—Weapon Detonations," 1975, pp. 6–7.

152 *Chemistry of the stratosphic clouds:* Richard S. Stolarski, "The Antarctic Ozone Hole," *Scientific American,* Vol. 258, No. 1 (January 1988), pp. 30–36.

152 *The role of air circulation and the polar vortex:* Richard S. Stolarski, "The Antarctic Ozone Hole," *Scientific American,* Vol. 258, No. 1 (January 1988), pp. 35–36.

152 *Evidence of ozone depletion in the other southern continents:* Reported in *Scientific American,* Vol. 261, No. 4 (October 1989), p. 26.

153 *Conditions in the Arctic:* Richard A. Kerr, "Arctic Ozone is Poised for a Fall," *Science,* Vol. 243 (February 24, 1989), pp. 1007–1008.

153 *Airborne research into the polar stratosphere:* Ellen Ruppel Shell, "Solo Flights into the Ozone Hole Reveal Its Causes," *Smithsonian* (February 1988), pp. 142–155.

154 *Concern caused by readings taken in the stratosphere:* Richard A. Kerr, "Arctic Ozone is Poised for a Fall," *Science,* Vol. 243 (February 24, 1989), p. 1008.

154 *1989 ozone hole:* Richard A. Kerr, "Ozone Hits Bottom Again," *Science,* Vol. 246 (October 20, 1989), p. 324.

154   *Solomon's comments on annual variation in depth of ozone hole:* Personal communication, October 1989.

154   *Measurements of ozone depletion over Northern Hemisphere:* William B. Grant, "Global Stratospheric Ozone and UVB Radiation," *Science,* Vol. 242 (November 25, 1988), Letters, p. 1111.

155   *Incidence of skin cancer:* "The Possible Impact of Fluorocarbons and Halocarbons on Ozone, May 1975," Interdepartmental Committee for Atmospheric Sciences, Federal Council for Science and Technology Policy Office, National Science Foundation, May, 1975, pp. 65–75.

155   *Projection of increased cases of skin cancer:* Estimates given in "The Possible Impact of Fluorocarbons and Halocarbons on Ozone," Interdepartmental Committee for Atmospheric Sciences, Federal Council for Science and Technology Policy Office, National Science Foundation, May, 1975, p. 56.

155   *1989 statistics on cases of melanoma: Cancer Facts and Figures—1989,* American Cancer Society, Atlanta, Georgia, 1989.

157   *Decline in photosynthesis caused by UV-B:* Leslie Roberts, "Does the Ozone Hole Threaten Antarctic Life?" *Science,* Vol. 244 (April 21, 1989), pp. 288–289.

158   *Effectiveness of CFCs as greenhouse gases:* J. Hansen *et. al.,* "Regional Greenhouse Climate Effects," *Coping with Climate Change,* Proceedings of the Second North American Conference on Preparing for Climate Change, December 6–8, 1989, Climate Institute, Washington, D.C.

158   *Pollution reducing UV-B:* William B. Grant, "Global Stratospheric Ozone and UVB Radiation," *Science* (November 25, 1988), Letters, p. 1111.

158   *Scientist's remark about pollution shield:* William B. Grant, "Global Stratospheric Ozone and UVB Radiation," *Science* (November 25, 1988), Letters, p. 1111.

160   *Damage from ozone exposure to animals and human beings:* George D. Clayton *et al.,* "Community Air Quality Guides, Ozone (photochemical oxidant)," *American Industrial Hygiene Association Journal* (May–June 1968), p. 300. Also U.S. Department of Health, Education and Welfare, *Air Quality Criteria for Photochemical Oxidants,* National Air Pollution Control Administration Publication No. AP-63, Washington D.C., 1970, pp. 8:33 and 10:5 to 10:8.

References 199

161 *Ozone injury to vegetation:* George D. Clayton *et al.*, "Community Air Quality Guides," p. 301.
161 *A study of ozone trends:* Reported by William B. Grant, "Global Stratospheric Ozone and UVB Radiation," *Science* (November 25, 1988), Letters, p. 1111.
161 *Control of Nitrogen Oxides:* Philip H. Abelson, "Rural and Urban Ozone," *Science* (September 23, 1988), Editorial 1569.

CHAPTER 8

169 *Banning CFCs will reduce warming:* Daniel Charles, "EPA's Plan for Global Greenhouse," *Science*, Vol. 243 (March 24, 1989), p. 1544.
169 *Massive reforestration:* Figures reported by *The New York Times*, "Scientists Suggest Planting New Trees," July 18, 1989, p. 21.
170 *EPA's plan for tree planting:* Eliot Marshall, "EPA's Plan for Cooling the Global Greenhouse," *Science*, Vol. 243 (March 24, 1989), pp. 1544–1545.
170 *Trees cleared in tropical forests: The New York Times*, July 18, 1989, p. 21.
170 *Debt-for-nature swap with Madagascar:* "U.S. Grant to Help Madagascan Forests," *Los Angeles Times*, Aug. 2, 1989.
172 *Energy and economic growth in the 70s:* Marc Reisner, "The Rise and Fall and Rise of Energy Conservation," *The Amicus Journal* (Spring 1987), p. 25.
173 *Energy usage in other industrial nations:* Marc Reisner, "The Rise and Fall and Rise of Energy Conservation," *The Amicus Journal* (Spring 1987), p. 27.
173 *Price of electricity rose 250 percent:* Marc Reisner, "The Rise and Fall and Rise of Energy Conservation," *The Amicus Journal* (Spring 1987), p. 26.
174 *Amana refrigerator:* Marc Reisner, "The Rise and Fall and Rise of Energy Conservation," *The Amicus Journal* (Spring 1987), p. 27.
174 *Savings by energy appliance efficiency:* Marc Reisner, "The Rise and Fall and Rise of Energy Conservation," *The Amicus Journal* (Spring 1987), p. 22.
174 *Carbon dioxide emitted by coal-burning power plant:* Data sup-

plied by Commonwealth Edison, September 21, 1989. Figures are approximate; different kinds of coal vary in thermal content and plants vary in efficiency.

175 *Lovins quotation:* Cited by Stephen H. Schneider and Randi Londer, *The Coevolution of Climate and Man.* San Francisco: Sierra Club Books, 1984, p. 320.

175 *Demand for oil:* Marc Reisner, "The Rise and Fall and Rise of Energy Conservation," *The Amicus Journal* (Spring 1987), p. 24.

176 *Lovins quotation:* Amory Lovins *et al, Least-Cost Energy: Solving the CO_2 Problem.* Andover: Brick House, 1981, p. 73.

176 *Alvin Weinberg's admission:* Cited by Stephen H. Schneider and Randi Londer, *The Coevolution of Climate and Man.* San Francisco: Sierra Club Books, 1984, p. 320.

178 *Relative costs of photovoltaic power and conventially generated electricity:* H. M. Hubbard, "Photovoltaics Today and Tomorrow," *Science,* Vol. 244 (April 21, 1989), p. 300.

178 *Reduction in solar cell costs:* H. M. Hubbard, "Photovoltaics Today and Tomorrow," *Science,* Vol. 244 (April 21, 1989), p. 300.

180 *Real-cost of gasoline compared to hydrogen fuel:* Robert Keating, "Road to Power," *Omni* (June 1989), p. 72.

185 *Whitman quotation:* Walt Whitman, "There Was a Child Went Forth," *Leaves of Grass.* New York: Norton, 1965, p. 364.

Suggestions for Further Reading

Kelley, Kevin W. (conceived and edited for the Association of Space Explorers). *The Home Planet*. Reading, Mass.: Addison-Wesley Publishing Company, 1988. A beautiful book containing many spectacular pictures of the Earth taken from space and comments by many of the early astronauts.

Francis, Peter, and Pat Jones. *Images of Earth*. Englewood, N.J.: Prentice Hall, 1984. More stunning photographs of the Earth (both natural and false color images) from the Shuttle and Sky-lab programs, with an explanatory text.

Schneider, Stephen H., and Randi Londer. *The Coevolution of Climate and Life*. San Francisco: Sierra Club Books, 1984. This book contains so many facts and theories about climate and living things that it almost serves as a course in itself. Written in a style that does not assume any previous scientific training.

Schneider, Stephen H., *Global Warming*. San Francisco: Sierra Club Books, 1989. A leading climatologist addresses the causes, effects, and impact of the greenhouse effect. An evenhanded treatment of the subject.

McKibben, Bill. *The End of Nature*. New York: Random House, 1989. A discussion of the global problems that are threatening the environment on Earth, with special emphasis on the greenhouse effect.

Pawlick, Thomas. *A Killing Rain: The Global Threat of Acid Precipitation*. San Francisco: Sierra Club Books, 1984. A good book for the general reader on the subject of acid precipitation.

Boyle, Robert H., and R. Alexander Boyle. *Acid Rain*. New York: Schocken Books, 1983. Another popular book on this subject.

Roan, Sharon L. *Ozone Crisis*. New York: John Wiley & Sons, 1989.
A science writer traces the evolution of scientific knowledge and
public consciousness concerning depletion of the ozone layer. A
detailed and very interesting account, it is written in language
that can be readily understood by the layman.

Lovins, Amory *et al*. *Least-Cost Energy: Solving the CO$_2$ Problem*.
Andover, Mass.: Brick House, 1981. This book contains much
detailed information about conservation.

Environmental Information Exchange Citizen Guide. *Protecting the
Ozone Layer: What You Can Do*. Available for $2 from the Environ-
mental Defense Fund, 257 Park Avenue South, New York, NY
10010.

For those readers who would like to pursue the scientific aspects of
these global problems in more depth, the following two issues of
scientific magazines give excellent general coverage: *Science,* Issues
in Atmospheric Sciences, February 10, 1989. And *Scientific American*, Special Issue: Managing Planet Earth, September 1989.

Index